STUDIES IN ANCIENT TECHNOLOGY

VOLUME II

STUDIES IN ANCIENT TECHNOLOGY

BY

R.J. FORBES

VOLUME II

WITH 38 FIGURES AND 7 TABLES

THIRD EDITION

E.J. BRILL
LEIDEN • NEW YORK • KÖLN
1993

This volume is a reprint of the second edition 1965.

The paper in this book meets the guidelines for permanence and durability of the Committee on Production Guidelines for Book Longevity of the Council on Library Resources.

Library of Congress Cataloging-in-Publication Data

Forbes, R. J. (Robert James), 1900-
Studies in ancient technology / by R. J. Forbes.—3rd ed.
p. cm.
Includes bibliographical references and indexes.
ISBN 9004006214 (v. 1: cloth).—ISBN 9004006222 (v. 2: cloth).
— ISBN 9004006230 (v. 3: cloth)
1. Technology—History. I. Title.
T16.F67 1993
609—dc20 92-39555
CIP

ISBN 90 04 00622 2

PRINTED IN MALTA BY INTERPRINT LIMITED

CONTENTS

PREFACE TO THE SECOND EDITION

The text of the first edition has been corrected and rewritten on several points and more recent literature has been added to the bibliographies. This has partly been made possible by helpful critics, the most outstanding of which was W. F. Leemans, who in his review of the first six volumes of this series made many valuable suggestions (J. E. S. H. O. vol. III, 1960, 2, 217—237) most of which have been incorporated in the present edition. We hope that others will follow this example to help the author gather information on the subjects discussed in this book.

Amsterdam R. J. Forbes

Chapter one

IRRIGATION AND DRAINAGE

Introduction

Galen reports Thales saying "Water is the substrate, and all things are derived from it." If we disregard the philosophy which Thales based on this hypothesis (and little do we know about it) it expresses perfectly the feelings of many primitive and ancient peoples, who make the world and all civilised life arise from the waters of chaos.

As water covers 72% of the globe, and seas, rivers, snow, hail, ice and rain have taken such an important part in the formation and deformation of the earth, water largely conditions and forms primitive life. Hence the overwhelming rôle water plays in mythology and theology, in ritual and magic (1).

Man and animal rely on water in their diet and for the preparation of food. Water was taken from brooks and wells (2), and the gradual development and mechanisation of these resources form the story of water supply. Secondly the use of rivers and seas as a medium of transport forms the story of the ship and shipping. Then water is used as a source of power; the development of the water-wheel started during the Roman Empire but reached its apex in modern times. Fourthly there were irrigation and drainage.

Irrigation is the artificial process of supplying water to crops in countries where the rainfall is insufficient or comes in the wrong season (3). Hence irrigation is practised only in climates with an annual rainfall of less than 15—17″ of rain. The old river civilisations of the Ancient Near East with their lack of rain and hot summers were the first to apply irrigation on a large scale. Irrigation requires a large amount of water. In distribution about 50% of the water used gets lost, whilst 25% of the water flows away unused, leaving but 25% of the water to be actually used by the crops, apart of course from the silt deposited (3) (4). Therefore large amounts of water are always needed for proper irrigation.

However, more is needed for irrigation. The geological structure of a country may determine the type of irrigation used. The Nile valley is a narrow strip of country between two desert plateaus. In Mesopotamia the Euphrates and the Tigris often wind their way above the level of a broad delta-valley. In the latter country the distribution and drainage of the water is extremely important. The quantity of water available from the perennially rising rivers, the season and the duration of the flood, the temperature and evaporation of the irrigation water, its silt, sand and salt content and their composition determine the methods used.

Drainage, the removal of excess water from swamps and other lands, such as those recovered from the sea or rivers, was more typical of the Mediterranean world, where the form of agriculture and the crops hardly ever demanded irrigation, and of Western Europe, where the rainfall was high enough to raise the crops. Hence irrigation was typical of the preclassical, drainage of the classical world.

Systems of irrigation

The simplest form of irrigation is lifting the water from the well or the river by means of a bucket, shadoof or other machinery (Table I). The evolution of this *water-raising machinery* to the steam pump or

TABLE I

Evolution of the water-wheel

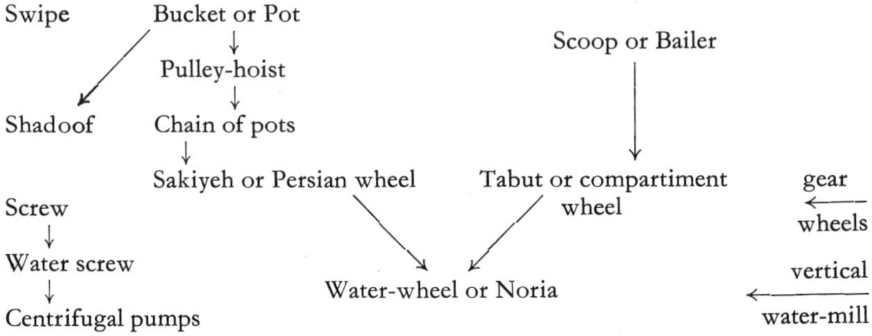

electric pump forms a separate story (page 32). It will, however, be recognised that in its earlier stages this form of irrigation would be of local use only and would hardly contribute to a permanent organisation on any scale except as the last link in such an organisation.

If providing for a larger area it might prove well worth the joint effort to construct a *canal* tapping the river upstreams and by a judicious choice of the slope of this canal this provides a means of "raising water" sufficiently high to irrigate a large area. The suppiy to the canal can be stabilized if a *weir* is built across the river just below the point where the canal takes off. Boths methods were used in Antiquity within the larger organisation of perennial or basin irrigation.

The ancient irrigation systems of the Near East were confined to the river valleys. Only in a few cases do we hear of irrigation practised in hilly countries such as Palestine or Armenia. The system chosen there was terraced hill-sides over which the water from brook or well was led. This system is so widely spread and so natural to all primitive peoples that we need not suppose any diffusion of such a system from any centre where it was invented.

In pre-classical times irrigation was mainly *natural irrigation* that is it depended on the natural rise and fall of the rivers, which had built up the plains of Egypt, Mesopotamia and the Indus valley. The rivers overflowing these slightly concave and sloping plains would carry water and silt over the fields. The heavy parts of this silt would settle close to the river banks and thus the rivers tended to build up their banks. The floods would thus naturally water less and less land and the silt would be carried out to the river's mouth building up a delta-land. Should the river-bed rise above the level of the plain, there is always a risk of the banks breaking and the river seeking a new bed.

Hence the primitive farmer has to correct and to supplement natural irrigation. He would have to strengthen and watch the natural dykes along the river. He might cut this dyke to conduct water towards his fields closing the gap when sufficient water and silt had been admitted. The excess water would be drained off by cutting the river bank downstreams. The system of *basin irrigation* is based on this simple scheme. The land along the river is divided into basins by dams perpendicular to the river's course. A canal conducts water from the river to each basin, where the water is divided by smaller canals and ditches. Another canal drains the excess water off to a second basin or to the river downstreams. Even this system with its set of basins and canals will not spread the water evenly, if the river has built up its banks too much and the flood water will not reach the higher grounds along its banks. In that case two strips of basins are constructed along the river, one comprising the higher, the other the lower-lying parts of the country. The lateral canal running between the two strips of basins bringing water to both, each having its own water level, is the final improvement over canals for each set of basins.

A chain of basins and a set of lateral and other canals form an irrigation unit, a "water province", naturally disposed to cooperation and regulation of the water level with the aid of gravity flow. Large quantities of water are needed. In order to flood the fields with 1—2 m of "life-bringing" water some 7,500—15,000 m^3 is needed per ha. The area of the basin varies from 1200—20,000 ha in present-day Egypt, but there are indications that they were fairly small in ancient Egypt and averaged some 2000 ha.

The system of *perennial irrigation* was the natural choice of ancient Mesopotamia where the river bed is high above the plains and the rivers can supply water even in the dry season. The farm land is cut by a great many canals, watering many fields of much smaller size. The main canals are often shut off from the smaller ones by dykes, weirs or locks. As the floods do not come in the correct season, the water is stowed in reservoirs, often natural depressions on the border of the desert (ancient Mesopotamia) or man-made earthen or masonry reservoirs (ancient India and Ceylon). The flood water thus preserved is led to the fields when wanted in smaller quantities (500—1,500 m^3 per ha) but now several times a year.

Table II will show under what conditions these competitive systems were used in Antiquity. The main advantages of basin irrigation are a short canal system, simple up-keep, little risk of salt accumulation and less speedy silting up through negligence; perennial irrigation helps to harvest crops more than once a year, but it means constant up-keep of a much longer canal system and the chance of salt accumulation. However, the larger and more frequent harvests were so tempting that the disadvantages did not at first weigh too heavily. Even in Egypt perennial irrigation was introduced in Hellenistic times and has ousted the older system completely now.

Irrigation and ancient society

Irrigation as applied to the sub-tropical river valleys must have enhanced the contrasts between the barren desert and the rich valleys. Generally speaking the valley soil contained salty and alkaline constituents which the irrigation water washed out and thus the floods meant a double boon. However, even the simplest form of adapted natural irrigation meant work. The digging of canals and ditches, the building of dykes and reservoirs meant handling large quantities of earth. This labour, unless on a very small scale, was impossible in a small community. Irrigation, even on a modest scale, meant cooper-

TABLE II

Comparison of the conditions of irrigation in ancient Egypt and Mesopotamia

	Egypt	Mesopotamia
Season of the floods	August to early October	April to early June
Climate	Semi-tropical	Continental
Average summer temperature	110° F	120° F
Average winter temperature	53° F	40° F
Season after the floods	Winter	Hot summer
Relation of harvest and floods	In time for winter- and summer-crops	Too late for winter-crops too early for summer-crops
Rise and fall of the waters	Slow and clear rise and fall	Sudden rise and fall
Profile rivervalley	Concave, sloping towards sea. No stagnant water	Very flat, faintly sloping towards sea. Pools and swamps
Surrounding country	Lime- and sand-stone hills	Weathered marls containing salt and gypsum
Type and Quantity of sediment	Sufficient, salt-free sediment. Little silting-up of canals	5 x as much salty sediment, canals silting-up quickly
System of irrigation	Basin irrigation	Perennial irrigation
Effects of irrigation	As result of irrigation and type of soil tendency to extract the salts present in the soil. Very slow silting-up of canals	Tendency of salts and alkaline compounds to accumulate in soil. Danger of silting-up of the many canals

ation and organisation of labour. Hence, it became possible only when a closely-knit social organisation and government worked for this government and because of it. The handling of large quantities of earth was possible because statute-labour (corvée) was a form in which taxes were paid. It demanded both control and differentiation of labour lest the work be finished in time.

Each special job had to be executed and timed in relation with similar ones in adjoining districts, with the rise of the rivers and the demands of the crops grown. This could only be done by a state or social organism which had the power and the officers to organize and control these jobs. Water was the object of many legal disputes and hence it contributed heavily to the evolution of common law.

Fig. 1.
Agricultural terraces of the Incas near Pisac, Urubamba Valley

Irrigation also stimulated science. As the fields had to be replotted after the inundation, the boundary stones set up again and checked, and as the taxes often depended on the rise of the waters and the area flooded, surveying the land was an important job. Geometry and surveying were both stimulated by irrigation. Also it was most important to determine both the time and the height of the inundation. Though primitive agriculture had already used astronomical phenomena to determine the seasons, irrigation stimulated astronomy, the evolution of the calender and the study of the rise and fall of the rivers.

Again irrigation stimulated engineering. In the first place it sponsored the evolution of water-raising machinery, and the application of cog-wheels in machinery in general. It also meant accumulation or handling large masses of earth, a skill much needed in ancient architecture which often erected high terraces and mounds to be faced with natural stone.

Irrigation and the plough greatly increased the harvests per acreage and thus provided food for a rapidly growing population. The earliest urban centres arose in the river valleys of the Near East. This concentration of population intensified trade with the mountain countries which bartered for corn their products such as building materials, ores, metals and timber. Thus irrigation and the evolution of urban centres are closely connected.

The decline of the ancient irrigation systems (5) is primarily one of failing central power. The silting-up of canals, lingering up-keep, salt accumulation in the farm lands, depopulation through heavy taxation, destruction of the canals by invaders were only secondary causes.

The beginnings of irrigation

Though it was at one time contended that irrigation was an invention of ancient Egypt (6) this opinion has no archaeological support and rested on misinterpretation of reports on the more favourable natural

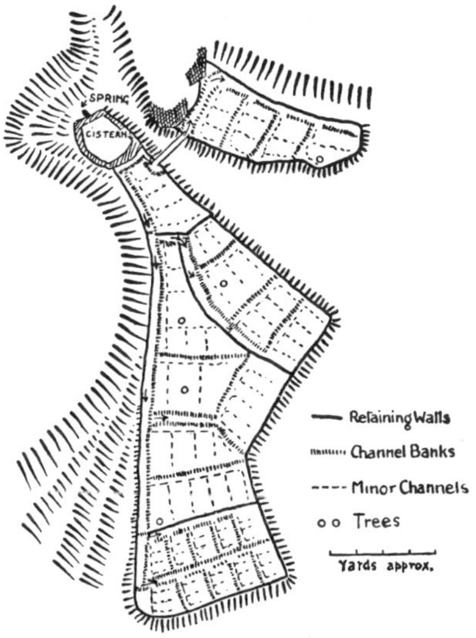

Fig. 2.
An irrigated Hopi garden at Sykiatki.
(After Daryl Forde, Habitat, Economy and Society, p. 232)

conditions in that country as compared with those in ancient Mesopotamia (7). Irrigation, as practised in the ancient civilisations, was the product of an evolution of many generations. The changeover from hoe-agriculture to plough and irrigation farming took place at different places on earth at various periods, its form conditioned by widely different physico-geographical conditions (8).

Ethnology (9) tells us of many primitive efforts at irrigation. The Kazaks of Siberia make weirs in the rivers during the dry summer to conduct their water on their fields. After a certain period the weirs are demolished again. The Payute of North America, too, build primitive weirs of timber and brushwood such as one also finds in southern Spain. The Hopi and Yuma Indians use the little water they get from wells and brooks to water their skillfully-built terraces. The spread of certain crops may be the reason of the introduction of irrigation. Thus terraces and irrigation were adopted by many when taro spread from south-west Asia west and east into Polynesia (10).

In South America the Incas and their predecessors built weirs and irrigated terraces. The extent of irrigation in Peru's coastal plain is quite comparable to that in Afrasia (11). There it was a conditio sine qua non of survival. The existing remains are undoubtedly pre-Spanish, when their extent was much larger than now. The more important of these irrigation systems must have been constructed and administrated by a body of irrigation officials directed by a centralised pre-Inca government. Irrigation as a system came from the highlands to the coastal area and initiated a fairly rapid rise of a more advanced society attested by many archaeological finds and probably connected with the early Mochica and Chimu civilisations (100 B.C.—600 A.D.).

The present-day Indians of the south-west use natural springs and waterholes, and conserve rain and storm water from the mountains and the high mesas in natural reservoirs by dams and other artificial modifications (12). However, in south central Arizona we find an impressive system of prehistoric irrigation (13), "a million-dollar project" constructed with nothing more than rude stone hoes and wooden digging sticks. In the Salt river valley there are some 125 miles of main irrigation canals, in the Gila river valley about half that amount. Many of these canals measure 30′ from crown to crown, and 10′ deep. There are several independent systems in the Salt river valley. The intake of the main canals is far enough upstreams to provide sufficient fall for the irrigation area. The lateral branches, some of which are now obliterated, measure many hundreds of miles. This system was built by the prehistoric Hohokam Indians who arrived about the first century A.D. and left the region about 1400 A.D. because of surplus water, alkalinity of the water-logged soil and dwindling productivity.

Even now we find similar attempts in the desert surrounding Egypt. Both the Hadendoa of the eastern desert and Arab nomads in Sinaï use the scarce rains to water the thin strata of silt in the wadis in which they sow corn. This water is often held and conducted by dykes and earthen walls constructed for the purpose. It is not impossible that similar primitive methods were used by the prehistoric Egyptians driven into the valley by the gradual dessication of the steppe highlands, now desert.

The need for regulation of the natural inundation arose when the change of climatic conditions during the Neolithicum drove the primitive farmers down into the river valleys which were not overgrown by a dense jungle. Archaeology teaches us that population, which grows but slowly during this period, starts to extend quickly towards the proto-historic period at the end of the New Stone Age, in Egypt during the Nagada II period, in Mesopotamia during the Uruk period. This growing population is very probably both the cause and effect of the drainage, cultivation and irrigation of the river valleys. It coincides with the rise of urban centres and larger social groupings needed for the organisation and cooperation needed for such efforts. Irrigation may have started as these valleys began to be populated, the large-scale systems which we associate with this word must date from the middle of the fourth millenium B.C. The first states in Egypt and Mesopotamia represent the amalgamation of the work already done by smaller tribes or clans and they crown the efforts of the Copper Age by joining the smaller irrigation units into large well-organised systems. Only in the Copper Age or proto-historic period do size of population, climate, development of agriculture and social organisation provide the factors needed for the building of stable efficient irrigation systems. We have no proof that irrigation was an invention of a private individual. More probably it was the final link in a chaim of many gropings and attempts to master the natural rise and fall of the rivers in the valleys.

Terrace irrigation in antiquity

To the dwellers of the desert ancient Palestine may have seemed a "land of promise", if compared with Egypt, Mesopotamia or even northern Syria it was but a "barren land" which needed its "springs of water". There certainly was some irrigation in ancient Palestine e.g. in the Negev, where such primitive irrigation remains were recently studied, and viticulture was found to have been practised (14), though on a small and local scale only for conditions did not favour it. In this hilly country the hot-dry summer (80—90° F.) is followed by a rainy season in which 28″ of rain fall from the end of October until Easter. The few springs and wells allow some local terrace irrigation. Agriculture in this country was gardening rather than large-scale raising of crops (15).

Hence we hear mainly of "well-watered gardens" the water of which is provided from wells with buckets and pots or by regulated brooks which flow between the narrow oblong fields watered every fortnight. In the south somewhat larger irrigation plots can be made, but in Samaria the fields are very small (16). Disputes over water seem only too well-known (17). The topography of Palestine induced its inhabitants to develop conduits bringing the supply of water from afar. Their cities tunnelled well-defended approaches to wells beyond the city bounds (18). Its typical form is the sinnōr, the tunnel and conduit bringing spring water to the bottom of a shaft in the middle of the hill-top town. However, the sinnōr did not serve irrigation, but water-supply. Nor could hand-dug wells ensure a proper supply of water to irrigate large areas of land. As proper scientific guidance by geomorphology, stratigraphy and other branches of geology made it impossible to spot large underground water-carrying strata, such digging of wells was usually hardly more than trying to trace down the source of the water seeping from a certain spot of the earth's surface. Nor was there any drilling of wells, though the Chinese have drilled for water and brine from at least the third century of our era. The oldest artesian wells were drilled in Artois (1126 AD) and in Modena, but only early in the nineteenth century were tidings of the sound Chinese drilling methods combined with applied geology to make drilling for water reasonably efficient (18a).

Reservoirs or "pools" were used both for household uses and for irrigation. We know not only many pools at Jerusalem but also at Gibeon (19), Hebron (20) and Jerus (21). Irrigation in Palestine could never compare with that in Syria where it was possible to build weirs in such rivers as the Orontes (near Homs) and the Kuwek (near Aleppo) (22). Irrigation in Palestine was always more dependent on rainfall like those many oases in the Northern steppes than on artificial canalisation of streams (23). In Roman Palestine there was a considerable development, dams were built and water was diverted over as large an acreage as possible. Both in Palestine and Transjordania artificial wells and cisterns were used, many of these reservoirs were destroyed in later centuries. Pilate "raised another disturbance by expanding that sacred treasure which is called Corban upon aqueducts, whereby he brought Hater (to Jerusalem) from the distance of four hundred furlongs" (24). wowever, such aqueducts like those in Syria (Damascus, Emesa,

Heliopolis, Antioch, Aleppo and Tyre) served the water-supply of the big cities rather than irrigation, but agriculture certainly profited from the general introduction of water-raising machinery in Roman Palestine.

Many of these aqueducts and reservoirs were built with forced labour but irrigation never reached the extent which it attained for instance in ancient Peru (25) where topographical conditions are even more difficult. Here, however, the strong central power of the Incas made an intricate terrace system watered by natural irrigation and a series of long conduits. About 1400 A.D. the Inca Viracocha had a 3.5 m wide, 3.5 m deep canal constructed, which led the water from Parcu to Cotapampa, 225 km away. The invading Spaniards were astonished to find all the hill-slopes covered with terraces, water from weirs in the brooks led over one terrace after another. Such a strong central power never existed in ancient Palestine.

In southern Arabia irrigated terraces were also built in Antiquity (26) the traces of which have been studied. Their influence was ever reflected in law. Water was collected behind big dams like that of Ma'rib, which Arabian tradition ascribes to Lokman. These terraces, similar to those depicted in Egyptian scenes from the "incense-country Punt", have caused many to locate the ancient Egyptian supply-centres for incense in southern Arabia (27). The famous Ma'rib dam was probably built by a Sabaean King about 750 B.C. [1]. In Western Arabia different types of cisterns were found each designed for a special purpose and until the early phases of Islam various combinations of canals, weirs and dams served to develop the production of wheat and dates on the larger estates in Hidjaz.

Irrigation in Armenia and Persia was conducted with the help of "qanāts", adits pierced into the hill sides to collect water from the gravel strata and lead it to the fields in open conduits (28). Tolman tells us that such qanāts were used both as water-supply lines (a city like Teheran gets its water from 36 qanāts) and for irrigation purposes.

He informs us that: "The art of kanat building is practiced by a guild whose members are known as *mukanni*, under whose direction the coolies perform the manual labor. The chief (the *mukanni bashi*) locates the exploratory shaft or shafts upslope from the desired portal of the proposed kanat with appropriate mysterious ceremonies to impress his clients, usually a company of land owners. After the exploratory shafts have located a satisfactory bed of water-bearing material, a series of construction shafts are sunk, taking advantage of the terrain. Work begins at the portal and construction shafts are sunk at varying distances; according to Fraser at maximum intervals of 100 yd., where the kanat is deep, and averaging 60 to the mile. The grade and direction of the kanat from shaft to shaft are determined by a crude surveying instrument, consisting of a plumb line with a heavy weight hung from a tripod or crossbar placed over the shaft. The plumb line is tied into a loop above the mouth of the shaft and a notched stick is inserted, spreading the looped portion apart. The sighting stick is adjusted to the desired grade and differences in elevation are recorded by knots in the line. The muck is carried by hand from the working face of the tunnel in a fold of burlap or goatskin and hoisted to the surface by windlass. There is no ventilation in the tunnels. Reflected sunlight has been used since ancient times. Accidents are frequent and the loss of life is great, but there are always others to carry on".

The Assyrian king Assurnasirpal II (883—859) used the principle of a qanāt when he "dug a canal from the Great Zab and called it Patihegalli (that which opens(?) for abundance). I caused the meadow tracts by the Tigris to be richly irrigated and planted gardens there". The canal mentioned by the king was an artificially-contrived arm of the Great Zab that began at a point six miles south of Nimrûd, opposite the modern town of Quwair, and supplied water to the area east of the city. The over-flow of the Zab, after having passed the weir, was carried through a tunnel driven through the rock at Negub like a qanāt from vertical shafts (28a) Sargon II found and admired them near Ulḫu, though he destroyed them (29). His son Sennacherib borrowed the idea to construct the irrigation systems of Niniveh and Arbela (about 700 B.C.). The qanāt, the ancestor of the modern aqueduct, is now used in an area reaching from Morocco and the Sahara to the confines of Turkestan. However, irrigation plays only a minor and local part in Armenia and Persia, which was extended somewhat in Hellenistic and Roman times.

Irrigation in ancient India

The Indus valley has been called "Mesopotamia between Sarasvati and Indus". The Punjab and Kashmir have an extensive system of canals which water an area of some 2,000,000 sq. miles. Three large canal feed a net of lateral canals some 4500 km long. We do not know whether this present system is identical to the ancient one which fed the cities of the Indus civilisation. It is certainly unlikely that the Aryans who invaded India about 1500 B.C. brought the art of irrigation to India (30). Though the excellent drainage and sewerage system of these Indus cities was thoroughly studied no further data on the irrigation system that supported them are known. The water level in the valley has since risen some 10—15′ and both the ancient river beds and the cities are deeply buried in silt. However, hardly any traces of ancient canals were found. This apparent lack of any preparation for dry seasons together with the representation of forest-loving animals on the seals seem to point to the view that the rainfall in Sind and the Punjab must have been much heavier (31). It seems, however, a misplaced argumentatio ex silentio to say "though there is no direct evidence that the water wheel was known to the people of Mohenjo-Daro and Harappa, the shape and make of these jars certainly suggest that they were used on water wheels; and this would of course explain why such a number were made and broken" (32).

The use of irrigation is certainly recognised in the Veda's (ca 1000 B.C.) and the Brāhmanas mention not only irrigation and the channels needed but add that manure should also be used to increase the crops. More specific details date from the Maurya dynasty (third century B.C.) of Chandragupta and his successors. They are confirmed by reports by classical authors. MEGASTHENES not only mentions the use of qanāts in irrigation but he tells us that there were special irrigation officials whom he calls by the Greek title "agronomoi" (33). Irrigation was a matter of great solicitude and naturally under the charge of the state which regulated the supply of water as it derived revenue therefrom. One of the tasks of the "agronomoi" was to survey the irrigation and land measurement. KAUTILYA, the Indian authority (34) on these matters mentions that the peasants pay in the form of taxes 20—25 % of the harvest, another 20 % on land "watered by hand", 25 % "if the water is carried", 33 % "if the water is pumped up" or 25 % "if the land is artificially inundated".

Of old brooks and rivers were dammed in India and water was tapped from the artificial lakes or reservoirs thus formed. If they are silted up a new weir is constructed in the river below or above the old one. This system first in use in Rajputana and Central India gradually spread southwards where there are now no less than 53,000 reservoirs in the Madras Presidency alone with a total of 30,000 miles of embankment. The building of such reservoirs was sanctioned by religion (35). One of Chandragupta's provincial governors built a weir near Girnar and thus formed an artificial lake called Sudarsana. This weir was repaired in 150 A.D. and again in 458 A.D.

Buddhism was an important factor in diffusing this care for irrigation throughout India. The edicts of King Açoka (250 B.C.) mentions special local authorities entrusted with the care for settlement, irrigation and surveying (36), and other texts speak of the "canal builders who conduct the water wherever they wish" (37). When the Sinhalese King Vijaya invaded Ceylon (504 B.C.) the invaders settled on the flat arid plains north of Dambulla which they rendered capable of supporting a considerable population by the construction of one of the most elaborate irrigation systems ever attempted by mankind (38). King Panduwaasa, Vijaya's successor built in 494 B.C. the Giant's Tank near Anuradhapura, the area of which (223 sq. miles) equals Lake Geneva. Its retaining dyke (or bund) is 15 miles long and 300′ wide at its floot. The huge tank at Kalaweva built in 459 A.D. is 40 miles in circumference, it is held by a 12-mile earthen dam with a stone spill-way. The construction of the 12—14 miles wide Padavil-Colon Tank and its 11 miles wide, 70′ high dam (300′ at the foot, 30′ at the top) occupied 10,000 men during five years. The so-called "sluices" in these Ceylonese tank systems are either spill-ways or holes in the stone embankments spilling the water into irrigation canals.

Water lifting seems to have been mechanised early in India. The Persian wheel or "wheel of pots" is mentioned as "well-machine" [2] or "water-machine-wheel" [3] [4].

Irrigation in ancient China

The hydrographic situation in ancient China resembled closely that in ancient Mesopotamia and that of the present Low Countries. The Hoangho or Yellow River and the Yangtze-Kiang wind their way tortuously through a valley built up of loess deposited by these rivers and their tributaries. A glance at Table III will show that during the inundation season these rivers become real torrents, or rather mud streams because of the enormous amounts of silt they carry down from the mountains (39). Often these floods break the dykes before they have even reached the ocean. Such catastrophes have often happened in Chinese history and the mouth of the Yellow River has been known to have changed several times. The river beds are usually many feet higher than the surrounding flat country and hence irrigation was not such a puzzle as drainage was, which latter could be accomplished only by the digging of long canals. Hence the struggle *against* the inundation floods dominates Chinese history.

TABLE III

Irrigation in the ancient centres of civilisation

	Egypt	Mesopotamia	Hoangho-valley	Indus-valley
Climate	Subtropical	Semi-continental	Continental	Subtropical
Rainfall per year	8—12″	8.5″	20″	10″
Average summer temp. Average winter temp.	110° F 53° F	120° F 40° F	80° F 23° F	115° F 60° F
Period of inundation	Aug. to early Oct.	April to early June	July to September	May to August
Season after inundation	winter	summer	winter	dry season
Origin of water	African lakes and Abessynia	Armenian Mts.	Kunlun Mts.	Hindu Kush & Himalaya
Increase in water quant.	4 x	8 x	16 x	3.5—4.0 x
Rise of river	5—7 m	5 m	4—7 m	4—5 m
Silt on 100,000 water	170	755	1000—2000	435
Type of silt	Clay with up-to 20% of sand	Loam with much lime	Loes	Very fine clay
Profile of river valley	Concave	Nearly flat	Flat	Faintly convex
Slope towards sea	1 : 13,000	1 : 26,000	1 : 35,000	1 : 7,000
Surrounding country	Sand- and limestone hills	Marl containing salt and gypsum	Loess-plain	Lime- and sandstone formation
System of irrigation	Basin	Perennial	Perennial	Perennial
Harvests per year	One	Two	Three in two years	Two

Organised irrigation began in protohistoric Lungshan times in China. Tradition tells us that the emperor Yau punished his minister Kun "because he wanted to change the natural course of things" when he planned to build dykes to retain the Yellow River in its bed. Kun's son Yü, the founder of the Shia-dynasty is praised by the Shu-king as the first large-scale builder of dykes, who also had the river bed dredged after a disastrous flood. MENCIUS (40) reports that he "had a humble dwelling only, but he expended largely on canals and ditches". It is certain that in the earliest historical period great attention was paid to the building of strong dykes and dams and that many canals were built not only to carry away excess water and thus drain swamps but also in order to ship grain more cheaply and profitably (4). We are told that in the feudal period the loess region contained many swamps in the large valley bottoms and that on the other hand fields were covered with saltpetre. One of the earliest canals between the Rivers King and Lo coped with both evils by draining part of the country and irrigating another part. Over 4 million acres were thus made available to the peasants.

The dykes and dams made at this early period had all the refinements of modern ones (42). Fascines and stone setts were used to protect the slopes of the dykes since the reign of the emperor Wu-Ti (ca 100 B.C.) (43). However, no proper remedy was found for a stable foundation of dams below the water level and therefore the catastrophes of the past could only be partly averted. As water could easily be tapped from the high river bed, the building of reservoirs and the use of perennial irrigation was easily introduced. Generally speaking Chinese agriculture always remained in the primitive stage of horticulture (44), the crops being grown from seed and transplanted in beds at a later stage. By intensifying this form of agriculture high yields were obtained at the cost of much labour.

The irrigated terraces of the province of Shantung certainly date back to the third century B.C. The oldest Chinese signs, the "bone characters", reveal that even during the Shang dynasty (1766—1122 B.C.) irrigation had not yet generally prevailed. During the second and third month of the year rain oracles, many of which were excavated, were still consulted to satisfy the ancient farmers. On the other hand the bone characters have complex signs such as "flowing water + field" and "water + rice"; this goes to prove that irrigation was already practised. During this dynasty a unit of farmland (about $1\frac{2}{3}$ ha) was created which consisted of nine equal plots, and roads and dykes were identical like in ancient Egypt.

The later "socialist" philosopher MENG K'O (MENCIUS, fl.c. 320 B.C.) tried to prove that this unit was reflected in the later sign for "well" which was supposed to represent eight equal plots grouped round a well or reservoir. Though he bases his state socialism on such hypotheses, they are now largely disproved. However, the signs of the classical Chinese script evolved from the pictographical bone characters during the Chou dynasty (1122—249 B.C.) show how irrigation has penetrated social life (45). We find such complex signs as "flowing + water" (canal), "soil + water" (swamp), "ditch + field + ditch + field + ditch" (frontier), "field + cross ditches" (farm plot), "water + unshackle" (irrigate) and the sign for weir later comes to means-"catastrophe"!

Historical records prove the care of the Chou emperors for irrigation. Special officers were entrusted with its care as the harvests should be as large as possible. Canals were built both for irrigation and drainage and for cheap grain transport. The Han dynasty (202 B.C.—220 A.D.)

Fig. 3.
Sargonid seal (Louvre A. 156) showing shadoof (2500 B.C.)

also lavished on irrigation and canal building which spread southwards with Chinese domination beyond the Yangtze-Kiang. Chang T'ang, minister of finance, started an elaborate system of canals in 115 B.C., thus connecting all the important production areas of ancient China.

As Chinese influence penetrated to the Lop-Nor region and the Tarim basin shortly before the turn of our era irrigation was introduced in these regions. However, no clear picture of irrigation of ancient China is possible, what is sorely needed is a publication of the Chinese sources such as volume IV. 3 of J. Needham's Science and Civilisation in China promises (45a).

Irrigation in ancient Mesopotamia

In Mesopotamia we find conditions much like those in China. The country has a continental climate, winter in Assyria was cold and stormy. It was less secluded but less independent than ancient Egypt. All the important raw materials had to be obtained from the mountain regions, even timber was scarce from the earliest times onwards. The country "between the two Rivers" was a wide alluvial treeless plain with many stagnant pools and swamps. Gradually the rivers with their muddy inundations have raised the plain and pushed their mouths forwards 2.5 km eastwards onto the Persian Gulf every century. Here too the dykes were often broken and the river beds found a new course. Cities like Sippar, Nippur, Ur and Uruk were originally built on the banks of Euphrates. Here too the river bed and the canals are above the level of the farmlands.

The classical authors have all praised the fertility of these plains. "Its corn is so abundant that it yields for the most part two hundred fold, and even three hundred fold when the harvest is best" (46) and THEOPHRASTUS (47) mentions a fifty- to a hundredfold harvest. STRABO (48) states that "the country produces larger crops of barley than any other country (bearing three hundredfold, they say)". However, regulation of the rivers and care for irrigation and drainage here again take the form of a struggle *against* the water like in ancient China.

The ancient inhabitants of this country were always strongly impressed by number and measure. They started to fight the hydrological chaos symbolized in their myths and epics as the "Flood", draining swamps and digging canals. Ea, the god of the ocean on which the earth floated (the "waters of the Deep (Apsû)"), Ningursu, god of the inundation, Tiamat and others were malicious powers which had to be propitiated with gifts and prayers, rather than praised like the Nile (Hapi) by the Egyptians.

The earliest al-'Obaid inhabitants of Southern Mesopotamia (Sumer and Akkad) have founded their huts on mats made of reeds and they drained their swamps with canals. The Tigris rises earlier than the Euphrates, it carries about 2.5 times as much water but its bed cuts deeper into the plains. Therefore the Euphrates though carrying only 40% the amount of water, is the main source for irrigation of the land between the two rivers, as its bed is above the plains. The Tigris was used in Antiquity to water the country on its left bank, as is still the case east of Bagdad. The fact that the rivers rise in the wrong season forced the earliest inhabitants to adopt perennial irrigation. The possibilities of irrigation in Mesopotamia have been overrated since Antiquity. "O thou that dwellest upon many waters, abundant in treasures", said Jeremiah (49) and modern authors (50) have drawn equally optimistic pictures. Accurate calculations, however, have shown that in winter 2,800,000 ha, in summer 1,200,000 ha might be irrigated at most (51).

Thus the ancient Sumerians constructed an intricate system of dykes, parallel and lateral canals, canals tapping water from the rivers, weirs and reservoirs to stow the accumulated waters and to release them in the correct season. Hence every year the dams leading from the main canals to the smaller ones had to be opened and closed again. This is correctly described by STRABO (52): "For the Euphrates rises to flood-tide at the beginning of summer beginning first to rise in the spring when the snows in Armenia melt; so that of necessity it forms lakes and deluges the ploughed lands, unless the excess of the stream, or the surface water, is distributed by means of trenches and canals. Now this is the origin of the canals; but there is need of much labour to keep them up, for the soil is so deep and soft and yielding that it is easily swept out by the streams, and the plains are laid bare, and the canals are easily filled and their mouths choked by silt; and thus it results again that the overflow of the waters, emptying into the plains near the sea, forms lakes and marshes and reed beds, which last supply reeds (for all kind of crafts)....

Now it is impossible, perhaps, to prevent overflows of this kind, but it is the part of good rulers to afford all possible aid. The aid required is this: to prevent most of the overflowing by means of dams, and to prevent the filling up effected by the silt, on the contrary, by keeping the canals cleared and the mouths opened up. Now the clearing of the canals is easy, but the building of dams requires the work of many hands; for, since the earth readily gives in and is soft, it does not support the silt that is brought upon it, but yields to the silt, and draws it on, along with itself, and makes the mouth hard to dam. And indeed there is also need of quick work in order to close the canals quickly and to prevent all the water from emptying out of them. For when they dry up in the summer they dry up the river too; and when the river is lowered it cannot supply the sluices(?) with water at the time needed most in summer, when the country is fiery hot and scorched; and it makes no difference whether the crops are submerged by the abundance of water, or are destroyed by thirst for water. At the same time, also, the voyages inland, with their many advantages, were always being thwarted by the two above-mentioned causes, and it was impossible to correct the trouble unless the mouths of the canals were quickly opened up and quickly closed, and unless the canals were regulated so that the water in them neither was excessive nor failed."

The fine silt was indeed not very suitable for large dykes which were sometimes strengthened by layers of reed mats, a method also adopted in the building of temple mounds. The destructibility of these dykes makes it very probable that none of these canals had a life of more than 1000 years (53). Air photography shows that there are many traces of old canals, 6—25 m wide, leading from the Euphrates downstreams or to the Tigris. We also have the names of scores of canals (54) but the topography of ancient Mesopotamia is by no means solved (55). The same holds good for the ancient reservoirs, many of which silted up quickly, when the water became "dead" as the Arabs have it. Nor do we know much about the actual site of many weirs. Some have supposed that the rapids of the Euphrates near Hit and Anah were used to tap river water and that the gradual corrosion of these rapids was the cause of the degeneration of irrigation (56). The construction of such weirs was certainly feasible at an early date in view of the experience in making dykes which sometimes assumed the size of mounds from which the word for mountain (šadū) may well have been derived (57). In many cases the actual site and size is reported in ancient texts.

We get the impression that the city-states which made up ancient Sumer were actually irrigation units or provinces (58) and that many of their wars were due to irrigation problems. The maintenance of agriculture but also the social fabric was based on the control of water, for there was and still is a sharp frontier between the desert and the sown. Grain tallies and their complementary marks of ownership form the basis of cuneiform script. The commercial theocracy of the city-state was based on agriculture and therefore on irrigation. Hence irrigation is echoed in religious poetry like the Tammuz hymns, and the construction of canals and dykes was the primary duty of the kings. Complaints that "those in authority seem to believe that their canals and public buildings are the only things that will interest posterity" (59) are unfounded and unjust. When the irrigation system was finally thoroughly destroyed on purpose by Hulagu and his Mongol hordes after the capture of Baghdad (1258 A.D.) this meant the final decline of Mesopotamia, though the weak central power had already led to insufficient upkeep of the canals and reduction of the area under cultivation. The archaic script of Mesopotamia already reflects irrigation

[5]. The early kings of the First Dynasty of Ur were very active. Ur-Nanše created many canals and reservoirs (59). Eannetum is said to have made a large reservoir called Lumma-dim-dug. Entemena of Lagash may have dug the large canal now called the Shatt el-Hai; Urukagina and he claim many canals.

During the Third Dynasty of Ur period (2100—1950 B.C.) and the subsequent Old-Babylonian period there was again a great revival of canal construction. Ur-Nammu, the founder of the dynasty left us an inscription enumerating the canals which he dug to promote the fertility of his land (60). There is no doubt that irrigation technique was well-understood by them (61). The texts mention that the lateral canals were closed by a "mouth", probably a weir or a spill-way. The "tail" of these canals was probably a reservoir, which is quite distinct from the special tanks of drinking water called "NAG-TAR". The work on these canals was of course compulsory labour. The area cultivated in southern Mesopotamia was probably some 30,000 km^2.

Though the kings of the Isin-Larsa period also claim efforts in this field (Sin-idinnam) and even left us contracts for canals (Rim-Sin), we have a third period of revival during the reign of King Ḫammurabi of Babylon (about 1700 B.C.) whose last nine years were practically wholly devoted to canal building. He started the Arakhtu canal leading from Babylon and later joined with the Royal Canal (Nahar malḫu), connected Euphrates and Tigris in the narrows and laid a canal down the Sippar. In his famous code of laws some deal with the enforcing of the upkeep of canals, the protection of tennants against inundations (§ 45) and guarantees for sufficient supplies of water (§ 48). This shows how strong an influence irrigation had on the formation and codification of law. His successor Šamšu-Iluna was also active and built a canal for Uruk.

Obviously most of the canals commemorated in the year-formulae of the kings were for irrigation purposes, their names often containing such elements as "which brings abundance". All these canals had to be dug by hand thus demanding large numbers of workmen. This labour-force is as a rule denoted by the general word for workmen (erim, *ṣabê*), sometimes they were hired, in other cases they were recruited from statute labour. The adjacent landowners were responsible for the upkeep of the smaller canals, some regulations on this point can be found in the Codex Hammurabi. Stipulations with regard to irrigation were sometimes made in deeds relating to landed property.

Various terms are used for canals according to their size, the larger ones being called *narû* (Sum. i_7) e.g. river, the smaller ones *palgu* or *atappu*, the latter term being used by Hammurabi's Codex. Various authors have recently studied this terminology (61a).

Irrigation was just as important to the western regions such as the kingdom of Mari. In the archives found in that city there were many tablets addressed to the king of Mari by Kibri-Dagan, the governor of the district of Terqa to the north of the town of Mari. These letters contain a number of technical terms and they give an impression of the difficulties and the complexity of the problems which confronted ancient technicians. Not only they had to secure the irrigation of the fields, but they had also to prevent the irrigation water from flooding towns and villages. The Ḫabur river, which crossed the district, was apparently provided with dykes (*erretu*), which broke or overflowed in times of exceptionally high water. If *balitum* means really a basin, it may be accepted that water was kept in times of high water in reservoirs in order to have water for irrigation in dry periods. Breeches in the dykes of such a reservoir, as occurred at the reservoir of Zurubbân, apparently near a town of that name, had to be repaired as soon as possible. Another important work was the clearing of the canals of reeds. Many people were employed on all these tasks, as is shown by the texts, especially when the work was urgent in times of inundation.

In Old-Babylonian times irrigation was just as important to the town of Susa. The canal (pa_5) from which a field drew water was as a rule mentioned in leases. One of the texts gives a list of canals together with the areas of the groups of fields irrigated by them, but plenty of further texts give names of canals too. An official named sukkallu seems to have supervized the canals, which had to be kept in gooed condition by the neighbouring landowners as in later historical periods. In certain cases, e.g. by presenting a special gift or tribute the donor could be extempted from being responsible for the upkeep of a canal.

There is a fourth revival of canal building in the Assyrian Empire. Nebuchadnezzar I (1146—1123 B.C.) could already curse his enemies "that Raman the god of sources and rivers fill thy canal with sand and leave thee a prey to famine". The Assyrian kings, however, built huge canals which brought the water from the mountains down to the plains east of the Tigris to supplement the ancient irrigation system using river water, though Assyria had a fair share of winter rains.

Assur-nasir-pal II (883—853 B.C.) and above all Sargon II (722—705 B.C.) worked to make Niniveh and the new royal city of Chorsabad beautiful garden cities with parks, well-watered fields and gardens which received their water from new canals and weirs. Sargon had learnt the secret of tapping water from subterranean strata during his campaigns against Ulḫu and the old mining country of Urartu (Armenia). He constructed the Nâr-Mi-Dandan near Bagdad, the "canal of the land of Umliaš" on the eastern bank of the Tigris towards the borders

Fig. 4.
King Zer of Egypt (2600 B.C.) opening the irrigation canals at the beginning of the inundation season. (Photo Ashmolean Museum).

of Elam (Persia) and reconstructed the canal of Borsippa to make it fit for shipping. His son Sennacherib (705—681 B.C.) undertook a vast scheme to irrigate the country around Niniveh east of the Tigris. It involved canalisation of part of the Khosr river, the building of weirs, canalisation of the brooks east of the Khosr river, regulation of the upper Atrush river, and the building of a 55 km aqueduct tapping the new Atrush above a new weir near Bavian and carrying its water to Niniveh (700—690 B.C.) (62). He also built a qanāt to supply Arbela with water (63). Assurhaddon (680—669 B.C.) brought the water of the Zab to Kalah through the Negub-tunnel (qanāt). The Neo-Babylonian kings (625—538 B.C.) such as Nabopolassar, Nebuchadnezzar and Nabonidus reconstructed and amplified the irrigation canals wouth of Babylon. The texts mention many specialists in this field [6].

Alexander the Great and his successors understood the importance of these works and took good care for their upkeep. As STRABO has it (64): "Aristobulus says that Alexander himself when he was sailing up the river and piloting the boat, inspected the canals and with his multitude of followers cleared them; and when he noticed that one canal, the one which stretched most directly towards the marshes and lay in front of Arabia, had a mouth most difficult to deal with and could not easily be stopped up because of the yielding and soft nature of the soil, he opened up another mouth, a new one, at a distance of thirty stadia from it, having selected a place with a rocky bottom, and that he diverted the stream to that place; and that in doing this he was taking forethought at the same time that Arabia should not be made utterly difficult to enter by the lakes or even by the marshes, since, on account of the abundance of water, that country was already taking the form of an island."

The canal STRABO describes is called Pallacopas by Arrian (65) who says that "for three months over ten thousand Assyrians were engaged on this task". The topography of the region was entirely different, even two centuries after Alexander the Euphrates and Tigris still had separate mouths (66). The formation of the Shatt-el-Arab started the formation of swamps in this region. With Hellenism more rational irrigation spread into Persia, Bactria, Sogdiana and Margiana, and in the fourth century B.C. it flourished in Mesopotamia and adjoining regions. In the later Seleucid and early Parthian period the upkeep of the irrigation system faltered, the end of the second and beginning of the first century B.C. was the darkest period. Still Phraates IV (37—32 B.C.) could again praise his governors for having restored the irrigation system. Trajan and Hadrian are known to have constructed several canals. Sometimes private individuals constrcted them, e.g. in the neighbourhood of Dura Europos. It was handed over fairly intact by the Sassanids to the Arab caliphs, but suffered its final decline by the ruthless destruction of Hulagu's armies.

However, destruction of the famous fertility of the Land of the Two Rivers came not only by the hand of man. Nature itself by means of the composition of the fertile mud ruined many parts of Mesopotamia. During the centuries the large amounts of salts (notably gypsum) carried down the mountains with the silt made agriculture impossible in certain flooded districts in the south and have prompted the peasants to move north. Natural flooding as in Egypt would have avoided this, but natural circumstances pushed ancient Mesopotamia towards basin-irrigation with insufficient drainage facilities, which led to this accumalation of salts in badly drained areas and destruction of its fertility. The danger of the absorption of salts by the fields was recognised early in Mesopotamian history. A report written during the first year of the reign of Urukagina and found in the temple of the goddess Bau, the spouse of Ningirsu mentions that part of the fields belonging to the temple had become infertile. Studies by Prof. Thorkild Jacobsen in the Diyala region have shown the progressive destruction of the fertility of the irrigated fields by the absorption of salts from the inundation waters. Modern engineers (66a) are now discussing the possibility of curing this poisoning of the soil in Lower Mesopotamia. Therefore Hulagu simply signed Nature's decree of infertility.

IRRIGATION OF ANCIENT EGYPT

"The activity of the (Egyptian) people in connection with the river goes so far as to conquer nature through diligence... Diligence has oftentimes, even when nature has failed, availed to bring about the watering of as much land even at the time of the smaller rises of the river as at the greater rises, that is, through the means of canals and embankments" (67) says STRABO. Hence the Egyptians conquered the land *with the help of* the river. Hapi, the Nile (in reality the inundation waters) was a most friendly god. Even the three seasons are closely connected with irrigation, they are called "Inundation", "Coming forth" and "Lack of water" [7].

Diligence made this small strip of farmland along the Nile feed some 7,000,000 people in Roman times and even export large quantities of grain to Rome. Now cotton has largely displaced wheat. Irrigation was also supreme in the ancient Egyptian's paradise (68) and the god Osiris is early identified with the fertile waters of the inundation. Hence the dead would assure the judges in the Underworld that "he has never diverted a canal or raised a dam (to divert his neighbour's water) nor tapped another's ditch". Amenophis, son of Hapu's curse "May he be excluded from the waters of inundation" was a most potent one.

The Nile does not flood all Egypt. Though the rise at the first cataract is still 15 m, it is no more than 6—8 m at the Delta. Hence the Egyptian distinguished the naturally inundated land from the artificially irrigated fields [8]. About August 15th the water floods the land through the breaks in the dykes and remains there 0.5—2.0 m high for some 6—8 weeks. The title [9] given to the vizier probably refers to the fact that he performed the ceremony of cutting the dykes as is illustrated by the mace head of the protohistoric King Zer (fig. 4). Then the river started to fall and the water was drained off the fields, the dykes were closed again. This inundation coming after the harvest when the scorching sun had crackled the earth's surface would stimulate the refertilization and the extraction of salts. Perennial irrigation would not have had these advantages (69).

From the earliest historical period we find a well-organised "department of irrigation" which was probably a separate ministry and not subject to the ministry of public works of which all famous architects in Egyptian history were heads. We often encounter "chief of irrigation" among the titles of high officials (70) [10]. One of the most important tasks of this department was the observation of the rise and fall of the Nile. For this purpose wells were dug, in which a measuring staff or scale was placed, and this well was then made to communicate with the Nile. These Nilometers (71) date back to protohistoric times for the very early annals on the so-called Palermo Stone give the highest mark of the Nile for every year in ells, palms and inches (72). This already struck the classical authors, for Diodor reports (73): "In their anxiety over the rise of the Nile, the kings had a Nilometer built at Memphis. Those who were entrusted with the inspection noted carefully the rise of the river and they sent messengers to the different towns to inform them how many ells and inches the river had risen and when the fall began. Thus the people was relieved of all anxiety about it. All thus knew, after a long period of observation, in advance how rich the harvest of the gifts of the earth would be." Pliny (74) adds that a good harvest was expected from a rise of 12—16 ells at Memphis.

Neither too little nor too much, for this might spell a famine of which we have many records (75).

STRABO mentions a second Nilometer at Elephantine near the first cataract (76) where the rise should be 12 ells over that at Memphis. The inundation coincided with the season so called and also with the helical rise of Sirius. It marked one of the calendars in use in Egypt. The land was completely flooded along the rivers and the villages remained isolated on small mounds (77) which to the ancient Egyptians looked like the earth emerging from the waters of chaos at the day of creation. Hence in the ancient script the road is represented by a dyke with water lilies on its slopes. Basin irrigation was certainly the oldest system introduced in Egypt and its "provinces" were certainly irrigation units [11] like the Dutch "waterschap" or "hoogheemraadschap".

Each of these provinces had its own central office, the "waterhouse" whose chief directed a set of "inspectors of the dykes", "chief of the canal workmen", "inspector of the forced labour", "watchers of the Nilometers", "inspectors of the inundation" and officials "who cut the dykes" (or "opened the dams"). They in their turn a host of slaves. Thus Ramses III (1200 B.C.) mentions "I appointed many workmen as watchers for the inundation administration" (78). There were special lawcourts ("water tribunals") to deal with conflicts about irrigation water. The units mentioned were about 12—80 km^2, that is about one tenth of the present ones.

After the inundation there was also much work for a second state department, the land registry office. It was very old, for even the Palermo Stone mentions the "measuring and counting of the lands" on several occasions. This was important as taxes on farmland were levied in relation to its surface and the depth of the inundation. Also the boundaries, which were sometimes obliterated by the mud, should be re-established. "I have counted and measured the fields for you" said Ramses III to his father (79). Hence surveying became the mother of geometry, a fact that the classical authors have already noticed (80) and which the mathematical papyrus Rhind proves.

Every two years the land registry office measured the fields in detail, an operation called "counting" (tnwt) (81). The smallest unit was the aroura (c. 2750 m^2), a square of 100 Egyptian ells. We have many pictures of the surveyors, their scribes and workmen (82).

The "forced labour" was a form of tax which the peasant paid after the inundation, they were fed by the state during this period. This labour was used in the months April-June and consisted of the digging and clearing of canals and ditches. The dykes (with slopes of 1 : 3, top width 4 m) were enforced with pickets and reed mats, the water tanks (some of which contained 5000—10,000 m^3 of water) were cleaned. The sluices, sometimes mentioned (83), like those in Babylonia were not our locks but probably spill-ways built up of planks fitting into the dykes or in groved stakes rammed into the dykes, and thus acting as temporary gates. The irrigation apparatus consisted of buckets carried on a yoke and of shadoofs. All speculations about early water-wheels seem rather doubtful (84).

The unification of Upper and Lower Egypt by Menes (3200 B.C.) had completed a long process which began in the protohistoric period. It served irrigation well for centralisation could now improve efficiency. Many texts give us general facts on irrigation and we hear that during the reign of king Zoser the inundation failed for seven years and famine was the result. It seems that the first drainage and cultivation in the Fayum took place during the Age of the Pyramid Builders (IIIrd and IVth dynasties).

Even in the troubled times of the IXth—XIth dynasties, when central power was weak the independent local princes maintained the efforts to improve irrigation. Thus prince Kheti II of Hierakonpolis (2125 B.C.) tells us of an irrigation canal he built at Siut (85). "I brought a gift (of water) for this city... I substituted a channel of ten cubits. I excavated for it upon the arable land. I equipped a gate for its (mouth?) ...I supplied water in the highland district, I made a water supply for this city of Middle Egypt in the highlands which had not seen water. I made the elevated land a swamp. I caused the water of the Nile to flood over the ancient landmarks. Every neighbour was supplied with water and every citizen had Nile water to his heart's desire."

Irrigation technique also enabled the ancient Egyptians to build shipping canals, notably to avoid the dangers of the first cataract in the South. King Mernere (2400 B.C.) made the first canal at Elephantine (86), which was afterwards widened and deepened by Sesostris III (1875 B.C.) (87). This important canal protected by fortresses commanded the trade with Nubia and the Negroes. Many later kings like Thotmes I and Thotmes III (ca 1500 B.C.) worked on it too (88). Then the Egyptian kings have at several periods attempted to construct a canal connecting the Nile and the Red Sea (89) which canal never served for a long time for the shallows and storms in the gulf of Suez were a great obstacle to the primitive sea-going ships of Egypt.

New plans for irrigation were again attempted by the kings of the

Twelfth Dynasty, when central power was strong. Amenemhat I when appointing Chnumhotep, a provincial governor, stresses specifically the care for irrigation. Amenemhat III is probably the prototype of the "Sesostris stories" which the classical authors report (90). He was believed to have dammed the entrance of the Fayum depression and to have turned it into a water reservoir to irrigate the district around Memphis (91). Recent research (92) has disproved these theories.

The Fayum has proved to be a natural depression formed by the Nile in post-Palaeolithic times. Later when the level of the Nile sank considerably the Fayum was disconnected from this water supply. The lake then slowly dried up to the so-called "20 m level" already before the Neolithic Age. In historic times there never was a high-level lake in the Fayum. The oases were not even permanently inhabited (93). After the last rainy interlude which terminated about 4000 B.C. those inhabitants of the desert who did not invade the Nile valley withdrew to higher grounds in the south-west and south-east where they remained and maintained themselves till about 2500—2000 B.C. There was a doubtful recrudescence of rainfall in the Fayum in Old Kingdom times and in the classical period the Mediterranean rainfall moved a little farther south for a time. The work of the Twelfth Dynasty seems to have been the desilting of the Joseph's canal (Bahr Jusuf), the branch of the Nile, that originally fed the Fayum. In the oasis itself the inundated area was increased may be by building a dam and spill-way to retain the water brought down from the Nile.

The invaders, the Hyksos, seem to have neglected irrigation canals though there still were "the weary ones, the dead on the dykes" (94), but the empire builders of the New Kingdom (after 1580 B.C.) again took a great interest in irrigation problems. Unfortunately, we dispose of too little data on the population, harvests and inundated area to calculate the effect of evolving irrigation on the national income of ancient Egypt as some have attempted (95). With the advent of Hellenism and the dynasty of the Ptolemies we are much better informed on details of the irrigation schemes (96). The first Ptolemies immediately start energetic work on the canals and undertake extension of the arable land in the Fayum by irrigation. In this district we have 66 Greek names of towns against only 48 older ones. Ptolemy I sponsors viticulture both in the Fayum and Delta, which meant that higher grounds had to be irrigated. Ptolemy II tries to cope with the evils of a "low Nile". The later Ptolemies (Philopator & Epiphanes) neglected irrigation, the pressure of taxes led to the scarcity of labour, gradual depopulation and neglect of the dykes and canals. Under their successors there was a shortage of available land as the limits of irrigation were reached, they went beyond the present ones (97). The emperor Augustus had to reorganize the administration of the irrigation. The cultivated area was 884,600 ha in Egypt itself, over 123,300 ha in the Fayum and over 1,402,600 ha in the Delta. The correct maximum extent is not easily established now that Roman canals are buried by the desert, e.g. near the ancient settlement of Socnopaei Nesus (Fayum).

The Ptolemaic extension of the Pharaonic system was mainly obtained by the improvement of irrigation, the drainage of marsh land and careful irrigation of the sandy and stony borders of the desert. Most of the latter lands have since been given up. There was extensive tree planting on the embankments. These trees were first raised in nurseries and then transplanted. The garden lands of the Fayum and the Delta were so located that perennial irrigation was possible. In these lands water was available on request as we know from the oath of the "sluice guards". The artificially irrigated land in many cases paid higher taxes and this would be possible if perennial irrigation really raised more than one crop. However, the new system was not fully exploited that way as wheat was grown and not cotton like to-day. The low lands south of lake Moeris in the Fayum suffered from defective drainage hence they were reserved for the production of papyrus and fowling. We also know that Ptolemaic engineers constructed storage basins in the Fayum.

Hellenistic papyri also furnish us with more personal details on the irrigation officials of the period. We have the correspondence of the "architecton" Cleon, superintendent of the Fayum irrigation about 260 B.C., who also surveyed the drainage of 2700 ha of marsh land for the minister of finance Apollonius, who had to finance 84 km of dykes to add this land to his estates (98). Then there was Theodorus, chief of the Fayum irrigation engineers in the reigns of Philadelphus and Eurgetes I (c. 250—220). They worked with a staff of engineers (hyparchitecton) some of which, like Petchonsis, were definitely Egyptians by birth. They regulated the flow of the water in the canals and were in charge of the "water guards" (hydropylakes) who served 4—5 months during the inundation. With the exception of a few gangs of criminals (thesmotai) they hardly ever used forced labour, but depended most on corvée labour which was paid.

The dams (choma) were either enclosing dykes (perichomata), lateral dykes (diachomata) or weirs and spill-ways (emblema). The use of reed mats, wattled fascines and pickets to strengthen dams was well-known, some of them, and certainly the weirs, were protected with layers of stones. There were different categories of canals: the Nile and the main canals (potamos), lateral canals (dioryx), secondary feeder canals (hydragogos, eisagogos), secondary drainage canals (exagogos) and minor ditches. The corvée lasted 7—10 days (in Roman times only 5). The mud in baskets was often transported on donkeys or mules. The finances available for this work were derived from the "naubion" or tax on garden lands supplemented with irregular sums from the treasury

The months of the inundation were especially devoted to the inspection of dams and dykes, private owners were held responsible for their upkeep. In some canals there were stone jetties every 2—3 m to prevent the dykes being washed away. The greater extension of Hellenistic irrigation was due to the general introduction of good iron tools in irrigation and agriculture which was almost tantamount to a revolution. Also the larger use of water-raising machinery about the costs, leases and repairs of which the papyri inform us contributed largely to this revolution. Shadoofs and water-wheels of all kinds were now in general use, the Archimedean screw or "snail" was very popular in the Delta. Accounts of the first century A.D. show that they worked 92—129 days a year.

We have an accurate description and a map of the reclamation work executed by Cleon for Apollonius (99). Other bills for the excavation of canals and dykes were found, eight contracts deal with the transport of 292,733 m^3 of earth, another one specifies a 47,830 m^3 shift (100). The regular staff of the irrigation service was only for the general upkeep, New projects were given to private firms or contractors (ergolabai) after study of their tender. Private landowners often had their own civil engineers (potamitai). The introduction of water-lifting machinery by private owners was stimulated by a state premium on their output; this was proved by receipts found on the spot. Reed marshes are cut and burnt at a premium for the landowners.

Roman interest in Egyptian irrigation increased as this country became the granary of Rome and the "anona" (corn supply for the poor) floated down the Nile to be shipped to Rome at Alexandria. A "low Nile" now also meant famine at Rome. Hence the Romans paid close attention to the rise of the Nile we possess Roman records on the Nilometer at Elephantine dating from the end of the second century A.D. (101).

The emperor Augustus not only reorganized the service but he had his soldiers clear out the canals and construct new ones. Vespasianus had big embankments constructed at Oxyrhynchus (Fayum) and so did Titus and Domitian. There was some neglect in the revolutionary disorders of the third century A.D. but Probus again put the soldiers at work and there was a temporary revival at the end of that century. The work was financed in Roman times from the "water tax" on cisterns and garden land, a further tax on acreage and crop (the harvest being calculated to be 4.5—10 fold). The cost of irrigation services was calculated at about 13 bronze drachmae per aroura of 2750 m^2.

It should be remembered that private landowners often looked after their own share of the upkeep and new constructions. The inspectors for the maintenance of dykes and canals could assess the dyke tax (chomatikon). There were also overseers for the flooding and draining of basins and overseers of the weirs and sluices, which were maintained by a tax for costs and repairs. Sluice gates were often timber constructions (102). Sometimes the offices of "overseer of irrigation" and "inspector of sowing" were combined. The superintendent of irrigation regulated the flow of water more particularly in the perennially irrigated lands and the Delta, in which drainage sometimes caused great difficulties. This "strategus" was required under oath to supply the requisite supply of water. He was also in charge of the water guards. By the third century A.D. we read of guilds of "river workmen" who scour and dredge the canals.

The final downfall of the Egyptian irrigation system came in late Roman and Byzantine times. The decline of central power and above all the oppressive taxes led to depopulation of the villages, shortage of labour and no money to scour the canals. The accumulation of salt in many districts, especially in the Delta were the visible results of gross neglect.

Methods of raising water

The history of the devices used in ancient irrigation is still largely shrouded in mystery. It is clear that such machines existed in the later stages of Antiquity and that they were developed from the simple ladle, gourd or vessel used to collect water from stream or pool. The machinery moved by beast or man was finally driven by force of the river alone (103), by water-wheels. Antiquity did not know the suction-pump nor did it apply Ktesibios' force pumps (104) to irrigation.

The different devices ranging from the swipe to the water-wheel were used in classical Antiquity side by side and remained in use upto the present day, thus making our enquiry into their evolution more difficult. This is clear from modern terminology pertaining to the shadoof and the sakiyeh (105). The terms for the parts of these machines are mostly Arabic, but some of them are clearly of Coptic, Latin, Greek, Persian or Turkish origin. The same holds good for the whole of the Arab world (106). Persians, Nabataeans, Byzantines, Greeks and Romans all contributed to the spread of this machinery from the ancient Near East to the confines of the Moslem world. Where such devices as shadoofs are still used in the Western world their names betray their origin from the classical or the Moslem civilisations.

If modern terminology presents little help, classical and preclassical terms are equally vague. Nor does archaeology present much proof for these devices belong to what MUMFORD called the "Eotechnic phase of technology". They were built of wood, rope, pottery and mud-bricks, easily destroyed and frequently re-built and re-used in countries like Egypt or Mesopotamia lacking timber (107).

Technically speaking there is hardly any doubt the evolution of these devices (fig. 7). The most primitive form is the piece of cloth, basket or bucket attached to two sets of ropes and handled by two men. Apart from this swipe we have the leather bag or pottery vessel used to scoop water from wells and the curved wooden scoop or bailer (as used to keep ships dry), both being handled by one man only. These types of bailers are still in use in the Arab world and though their lift is only small they are well built to irrigate the small field slightly above the level of the last irrigation ditch. This is probably the method referred to in Ptolemaic papyri as "diacheiros". We have no direct earlier references to them though we know Egyptian reliefs showing the watering of gardens from pots and buckets. The SÚN, martabu, can not be a noria for reasons given further on. From the early pictograms given by FALKENSTEIN it would appear to be a bucket or a bailer with a handle. The latter form might explain the meaning "plough" as the bailer somewhat ressembles a primitive plough. Sumerian and Accadian texts usually remain very vague and mention nothing more definite then "watering the fields", but HERODOTUS (I.193) confirms the use of this method.

Where water was drawn from wells it was an easy step to use the camel or the ox running down a slope away from the well to draw up the bucket. With very simple means this vessel could be made to unload automatically at the mouth of the well into a gutter leading to the field.

Thus mechanisation was soon achieved but the output was limited by the depth of the well. We have few early records of this device so common now in the Arab world (108).

However, drawing water with a rope and vessel could also be mechanised in a different way. One could attach the bucket and its rope or chain to one arm of a balance and counter-balance its weight either by a fixed counterweight or by working a nooze in a rope or a

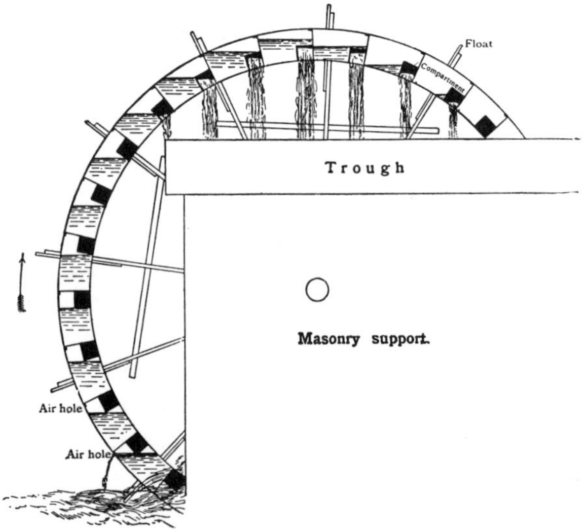

Fig. 5.
Sketch diagram of Tâbût. Section of wheel to show water in compartments and principle.

pedal attached to that arm by foot, that is by weight of one's own body. This is the principle of the well-known shadoof, a device that certainly goes back to early historic times like the balance. Spiegelberg (109) believed that he had identified the Egyptian term for this instrument but his conclusions do not seem definite.

Older literature always maintained that the shadoof was introduced into Mesopotamia from Egypt after Sennacherib's conquest (e.g. ca 700 B.C.) (110) but this can not be true as we find a shadoof on a seal-cylinder of the archaic period (fig. 3) (111). It would also seem that the well-known collection of tablets "ana ittisu" refers to an irrigation device made of wooden parts easily put together, "^{giš}lal = delutum" which may well be our shadoof (Taf. IV. col. II, 23—35). Possibly the fact that Mesopotamia was more generally suitable for irrigation by gravity flow, and Egypt had relatively more land that had to be artificially irrigated (the "gé abrochos" of the Hellenistic papyri) may be the proper explanation for the relative scarcity of the shadoof in texts from Mesopotamia.

Whereas most seals or pictures show the shadoof with a counterweight, the type moved by the pedal and foot was equally simple and

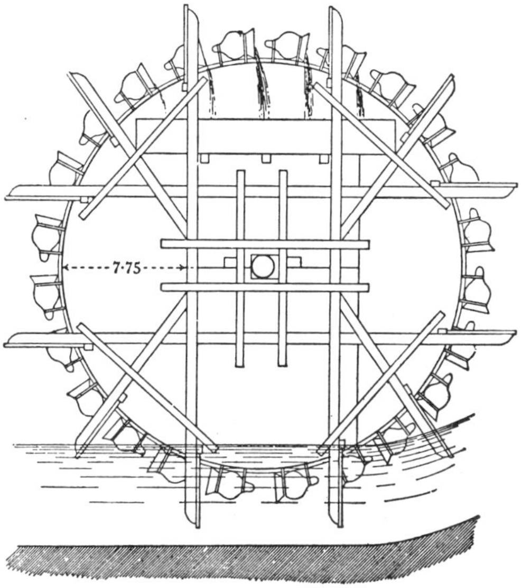

Fig. 6.
Wheel of pots for raising water. Scale $^1/_{50}$.
From WILLCOCK's *Egyptian Irrigation*

probably equally old. It is not described by PHILON but it is referred to in Deut. 11.10 "Thou wateredst the land of Egypt with thy foot" and introduced into Palestine in the Hellenistic period. It is well known in the classical world and both types survived in Western Europe and Asia upto the present day.

The next step, technologically speaking, was a fuller mechanisation of the drawing of water by bucket or scoop. Attaching a row of pots to the rim of a revolving wheel dipping in the irrigation canal one could raise the water to nearly the full height of that wheel. This is the wheel of pots or Persian wheel. From the scoop we derive the wheel with hollow compartiments in its rim, the compartiment wheel or tympanum. Both forms were moved by gearing a horizontal wheel turned by man or beast to the vertical water-hoist. Hence their introduction can not antedate the invention of geared machinery. The invention of the toothed wheel is sometimes ascribed to Archimedes and as far as our present evidence goes it certainly can not be very much older than 500 B.C. The only rotary devices antedating the fourth century B.C. are the potter's wheel, the vehicular wheel, the pulley (700 B.C.), the bow drill, the bow lathe and the reel and spindle of the weaver.

The possible explanation may be found in the widely different outlook of life of the pre-classical and the classical world and above all in the rising tendency of the Greek philosophers to apply natural phenomena and their laws to daily life. Pre-classical (mainly Babylonian) astronomers certainly watched the sky and observed stars and planets but they abstracted their observations by submitting them to mathematical (mainly arithmetical) treatment in order to wrench from them the laws of periodicity of celestial phenomena in which they were so interested. The Greeks, however, cast their observations into a geometrical scheme explaining the movement of the celestial bodies. Thus they created a system reproducing the mechanism of the heavens by complex rotary movements, the very atmosphere in which circular movements in all their aspects were studied and could be applied to other fields.

This is what VITRUVIUS expresses so well when he says (112): "Now all machinery is generated by Nature and the revolution of the universe guides and controls... Since then our fathers had observed this to be so, they took precedents from Nature; imitating them, and led on by what is divine, they developed the comforts of life by their inventions. And so they rendered some things more convenient by machines and their revolutions and other things by handy implements."

It would, therefore, seem that the Greek philosophers, in studying the mechanism of the heavens and in solving such mathematical problems as the division of the circle into a given number of equal parts, promoted rotary machines and gear wheels as a side line. Though the philosophers were not interested in these developments but theoretically, the craftsmen were quick to use them and our texts and archaeological finds confirm this.

On the other hand it is quite possible that such machinery is older than we believe on the strength of the present evidence from excavations. Laessøe who studied the literary evidence from cuneiform texts believes to have found evidence that the term *maialtum* denotes an irrigation machine moved by oxen, and he has argued that the shâdûf, the čerd and possibly the sâqia too go back to an early period in Mesopotamian history (113).

In Egypt these rotary devices were centainly introduced slowly in the Hellenistic era. In Mesopotamia they certainly have been used somewhat earlier if we take the evidence on the hanging gardens of Babylon. STRABO (113) tells us that "by the side of the stairs are water-engines by means of which persons, appointed expressily for the purpose, are continually employed in raising water from the Euphrates into the garden." This is confirmed by DIODOR (114) who says: "One arch had in it certain engines, whereby it drew plenty of water out of the river through certain conduits and contrivances from the platform."

These stories were confirmed by the excavations of Babylon conducted by KOLDEWEY. The three shafts found by him can hardly have contained shadoofs, nor were they suitable for wheels of pots geared to a horizontal wheel driven by man-power. It would seem that they contained chains of pots, that is technically speaking an endless rope with buckets attached to it and moved by turning the horizontal axle onto which the rope is attached. Unfortunately no remains testify of the actual construction of the chain of pots which were no doubt used in the hanging gardens of Babylon round 700 B.C. However, they must have been of the simple type described by PHILON of Byzantium about 200 B.C. and still common in certain parts.

We have already argued that the development of the Greek sciences of astronomy and mechanics in the fourth and third centuries B.C. stimulated further mechanisation of irrigation devices. Their gradual penetration is evident when we turn to the Hellenistic papyri found in Egypt. The sakiyeh is first mentioned in the second century B.C. (115) and then in 5 B.C. (116), it becomes much more frequent in Roman times. It should be noted that the very important Zenon papyri mention neither the sakiyeh nor the water screw. This agrees with the fact that VITRUV "begins to explain, so that they may be known, machines which are rarely employed" (X.1). He then describes the tympanum (X.4), the wheel of pots (for greater lifts) and the chain of pots (for maximum lift). This machinery which became more general at the beginning of the Christian era was driven by man or beast. VITRUV's tympanum driven by a treadmill provides an "abundant supply of water for irrigating gardens or diluting salt in salt pits" (X.4).

We have very few definite indications of the use of the tympanum or compartiment wheel in the Ancient Near East (Fayum c. 20 B.C.)

TABLE IV

The ancient forms for machinery used for irrigation purposes

A. (Intermittant) *Discontinuous Water Supply* from streams or canals

a. *By hand.* Max. lift : ca 1 m.

1.1 *Basket or bag moved by two sets of ropes* (two men!) Swipe, Wasserschleuder, Arab. nattal, naṭaleh, quffa; Greek "dia cheiros" (Greek Papyri).

1.2 *Leather bag,* Arab. dalu; Aram. delu (Num. 24.7); Hebr. deli (Num. 24.7), (Jos. 40.15), (John. 4.11).
Also used in wells.

2. *Curved wooden bailer or scoop with handle (one man)* Eg. ikn (I.139.18), pnḳ (I.511.1), mdəb.t (II.188.13).
Also used for ships.

b. From wells by animal power (ox, camel).
Well = Hebr. ʿen, bêʿer; Arab. bir, bêʿor.

1. Leather bag or wooden vessel, buckets, gourd.

2. Earthen ware jars.
Hebr. kad (Gen 24.13; 24.43; I Ki 18.34; Eccl. 12.6); Arab. garra; Syriac. mānā (= apparatus); Greek hydria (John 4.28) or keramion (Marc. 14.13).

3. Stone vessels.
Greek "lithinai hydriai" (John 2.6).

4. Well drawn by animal on draught-plane, Arab. gelib (Yemen), ġird; mhote (India).
Irrigates 0.4—1.5 ha.

5. Self-emptying pulley-hoist.
Arab. ġisd (Iraq), ġird (Arabia), kalabai (India);
Persia: cherad.

B. *Semi-mechanical.* "*The balanced bucket*". Max. lift ca. 1½ m.

1. *With counterweight*
The shadoof or swape
Arab. šādûf (Egypt), dāliya (Iraq), picottah (Malabar), lāt (India); Greek: keleneion, kelenion (Herod. I.193; VI.119), géranos; Egypt......?; Hebr. kīlōn (Mishna); Akkad. delutum (^{giš}lal); Latin: tolleno (tollo; toraise) (Pliny 19.20); French: Cigogne.

2. *Balanced by (weight of) the human body*

	Lift	Capacity	Acreage
Single step shadoof	1½ M	2700 l	0.4—1.5 ha
Three steps	4 m	1650 l	0.4—1.0 ha

Arab. dāliya (the scooper), ḥoṭṭāra (Souf) ḫaṭṭāra; Latin: tolleno (Philon).

C. *Continuous water-supply* by water lifting machinery.
Greek: organon, mechanè.

a. Moved by animals or men, Max. lift 1 m.

1. *Archimedean screw*, *water screw*.
Greek helix, cochlias (organon); Latin cochlea, vis Archimedis; Arab. ṭanbāra (Upper Egypt); French limace, escargot d'eau.

2. *Wheel of pots*, *Persian wheel* (Bucket-wheel hoist).
Greek trochos, mechane antlousa; Aram. galgelā deanṭelaya; Hebr. 'ōddet el-bayāra; anṭleja, from Greek antleon ("Emptier, bailer"); Arab. saqiya (the "irrigator") (Egypt), sâniya (Morocco); Spanish acena, cenia; Sanskrit kûpa - yantra ("well-machine").

3. *Compartiment wheel* (Hollow-wheel water-hoist).
Arab. tābūt (Egypt); Greek Tympanon; Latin typanum.

4. *Chain of pots* (without gears) (Philo of Byzantinum, 200 B.C.).
General terms: Arab. 'daulâb" (camel-wheel), haṭṭāra (HṬ-R = moving to and fro).
Persian dūlāb, Min. lift 4—5 m; Acreage 2—5 ha.

b. Moved by "the force of the river alone" (Vitruv. X:5). Waterdriven water-wheel (evolved after introduction of the water-mill).
Greek organon (II-VII AD) trochos, antletikon (antleo = to empty, bail), kykleuterion (VI & VII AD) mechanikon organon; Latin rota, organum; Pali: arghaṭṭa; Sanskrit jala-yantra cakra ("water-machine-wheel"); Arab. manganūn (Greek magganon), nā'ûra (N = R: making blood flow, compare Aram. R_1 â R_2 ū R_3 for instruments), noria ḥamâna, garb; Hebr. mukhnī (= mechané) "in the court of the Temple" (Talmud).

despite VITRUVIUS' elaborate description. Usually the evidence on these machines is fairly vague. STRABO tells us of the ridge of fertile land of Babylon in Egypt" on which water is conducted up from the river by wheels (trochoi) and screws (cochliai) and 150 prisoners are employed in the work" (117). Here again the machines are not moved by the current of a river (water-wheel) but by a geared treadmill.

The many terms mentioned in Hellenistic papyri are mostly far from clear (118). The "trochos" is made of wood and encased in masonry, it is identified with the sakiyeh. The general term for these devices is "mechanè", the word "organon" is also used and later transferred to the machines moved by waterwheels like many of the other terms. This often makes proper identification very difficult. The wooden "keleneion" or "kylon" is always the shadoof, however (see Table IV).

The papyri mention specialists to operate each of these machines, enlarge on the lease-money, hire and taxes to be paid for them, etc. In the second century A.D. 18 ob. was paid for the lease of a sakiyeh (119) which price should be compared with the irrigation costs per aroura of some 13 bronze drachmae.

In Palestine where irrigation was dependent on rainfall rather than on artificial canalisation and water-supply the wheel of pots and such machinery were introduced in Hellenistic times too. They were introduced into Mesopotamia in the same period and thence passed further east to India where the Persian wheel comes certainly later than 50 A.D. like other rotary devices.

The invention of the Archimedean screw or water screw, the last of these mechanical devices, can be traced with some certainty. DIODOR (120) claims that the Delta was irrigated with "mechaneis" invented by Archimedes of Syracuse and, because of their form, called "snail" (cochlias). In a further passage (121) he says that they were already in use in mines and wells before Archimedes introduced them in Egypt during his visit to the court of the Ptolemies. If Archimedes is not the real inventor himself the water-screw, which is nothing but a wooden screw turning in a hollow cylinder (122) can not have been devised but after the invention of the screw proper. This important device is ascribed by Greek tradition to Archytas of Tarentum who died in 394 B.C. Both Archytes and Archimedes had the mathematical knowledge needed to invent and construct screws and we know that Heron already possessed screw-cutting devices. Hence the classical tradition about the "vis Archimedis" seems virtually true.

Though the water-screw is still common in certain parts of Upper Egypt for small lifts and likewise in other parts of the Arab world it has now disappeared in the Delta region. The Zenon papyri do not mention it and it figures rarely in other collections of papyri until the Roman era. In Morocco it was introduced in Byzantine times and thence spread to Spain and France.

From the remains of an actual ancient water-crew found in Egypt (123), from pottery models and from the description by VITRUV (X.6) it is clear that they were worked by foot in Antiquity and not like some more modern types with the help of a crank, a device unknown to ancient engineers. PLINY propagated the use of water-screws for irrigating gardens and we have already quoted STRABO's evidence on the use of "cochliai" in Egypt (c. 807).

This brings us to the last stage of the evolution of water-raising, the fully mechanised water-hoists moved by a water-wheel. Screws and other water-raising machinery were used in Roman mines (123a). Here again we have fairly full evidence both literary and archaeological, to date this last step. It implies the knowledge of the water-wheel as a prime-mover. The most primitive water-wheel (the horizontal wheel vertical shaft water-mill), commonly called the Greek or Norse water-mill, is fairly old and it probably originated in the mountain region of the Ancient Near East to spread to the East and the West in the first century B.C. or even earlier. Though it was suitable for moving mill-stones and thence used for small scale corn-grinding, its vertical shaft would make it unsuitable for moving waterhoist such as we described above except with elaborate gearwheels. When the Roman engineers of the first century B.C. devised the modern water-wheel (the vertical wheel horizontal shaft watermill so clearly described by VITRUV X.5) "driven by the force of river alone" they created the proper prime mover which could be coupled directly on the same shaft to the water-raising machinery already in use.

The water-mill was definitely primarily a corn-mill. As such it developed slowly during the Roman Empire, then its use expanded rapidly in the West between the fourth and tenth century A.D. and to the East it came with Buddhism. In all those areas the waterwheel was first used to grind corn and only later it was applied to riase or pump water.

Hence these fully mechanised water-wheels appear later than the sakiyeh, the tābut or the water-screw. In Egypt they definitely belong to the Roman period (124) (the water-wheel of Arsinoë of A.D. 113). The papyri mention the cost of the brickwork of a water-wheel, which in A.D. 136 equals 2000 drachmae (125). In the next year a new water-wheel including the brick-work alongside the wheel on which the shaft rests is erected for 3000 drachmae (126). In A.D. 144 the rent of a water-wheel in the Fayyum is fixed at 140 drachmae (127). In Palestine too these water-wheels become common in the Roman period only, the Talmud mentions one in the court of the temple (128). We have no

Hellenistic texts on the water-wheels of the Tigris and Euphrates but in the Arab world they are quite common there.

It should be realised that most of these water-wheels, like the water-mills of Antiquity, were undershot-water wheels, which depended on swiftly flowing water and a fairly constant volume of water the whole year through to work efficiently. The overshot water-wheel, though

TABLE V

Evolution of the water-wheel

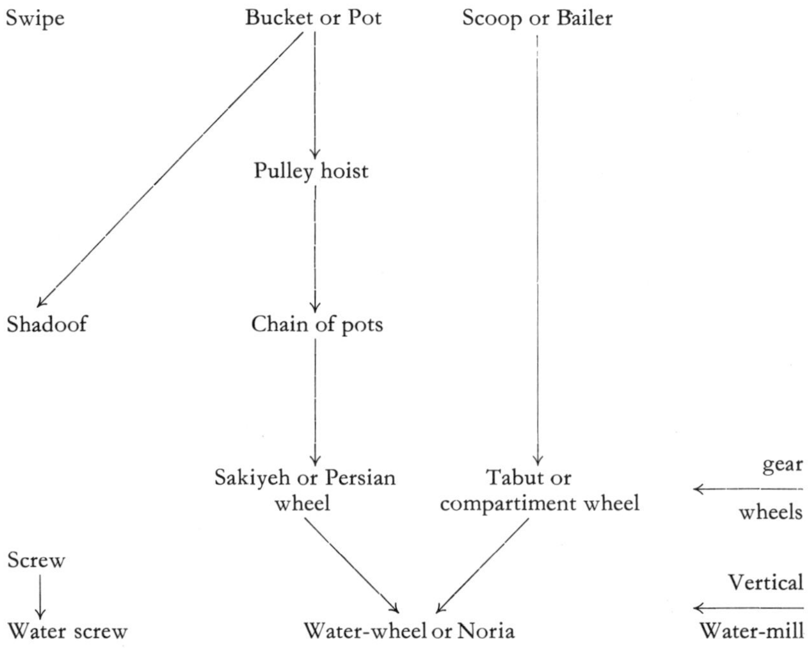

technically more efficient needed a constant supply of water from an aqueduct or if fed by sluggish rivers in a flat country demanded the construction of a millrace, a mill-pond and a chute plus sluice for proper manipulation. These conditions were rarely fulfilled in Antiquity and except on the banks of great rivers like the Nile, the Euphrates and the Tigris water-wheels remained rare and expensive machines even for such important uses as irrigation. With the advent of Islam and the rebirth of Hellenistic mechanical science came a renaissance of the application of all the older types of machinery to irrigation (129). But this belongs to the story of the water-mill as it was rather a quantitative and not a qualitative rebirth of the water-wheel and its predecessors.

The slow increase of the number of water-wheels under the Roman Empire is reflected by the birth of such terms as "antletikon" (a general term for irrigation machinery combining a water-mill and a water-hoist), "ergates", "kykleuterion" (a term of the sixth century A.D. denoting machinery driven by a water-mill) and "organon" (used from the second upto the seventh century to denote all machines connected with a "lakkos" a well or water cistern). The Arab terms, many of which have a Greek origin, clearly denote their Hellenistic or Byzantine parentage.

The noria not only spread to the west (Morocco, Spain) but also to Asia and the Far East. According to Japanese tradition the prince Yoshimine Ysuyo (785—830) introduced the noria in Japan for the irrigation of the rice-fields.

In the water-supply systems of Imperial Rome the water-wheel plays an important part in obtaining a good head of water when the supply was delivered at a low level (130). The main baths of Ostia, the port of Rome at the mouth of the river, have storage tanks from which the water was lifted by a wheel to an upper-floor tank, whence an adequate supply at fair pressure was obtainable, for at Ostia the supply arrived at a low level under the streets. A similar method was adopted in Roman Silchester and other sites. It remained in vogue in Europe until the last century.

Irrigation and drainage in the classical world

In the classical world irrigation was not a complement but rather a supplement of farming. Water conservation was the principle at work in the classical technique of cultivation, based on the winter rains, the water of which had to be conserved in some way as in the months June, July and August less than 4″ of rain fell. The Mediterranean region consists of granite and crystalline plateaux alternating with folding mountain chains. Most of the mountain slopes suffered from denudation every rainy season when torrential streams tear down. the detritus, and this grew worse as the forests were cut down. There was too little manure available to supplement the winter rains. It was usage to have an alternate fallow year and cross-ploughing, which demanded a lot of cheap labour, was in general use.

The seasonal rainfall was irregular and torrential. To avoid evaporation the soil was pulverized mainly with hoes and rakes, drainage was very important. The water percolated very slowly, drawing masses of humus with it; the weathering of the soil penetrated to much greater depths than in the North.

This "dry farming" technique propagated by the mass of classical agricultural doctrine held for Greece and Italy, but failed in the wetter climate of Western Europe. It was an accumulation of experience upon relations of crops and soils. "It is the weather rather than the soil that determines the harvest", said THEOPHRASTUS. The same technique was applied to a variety of vegetable and fodder crops, which could also serve as green manure. The same technique was often applied to olives,

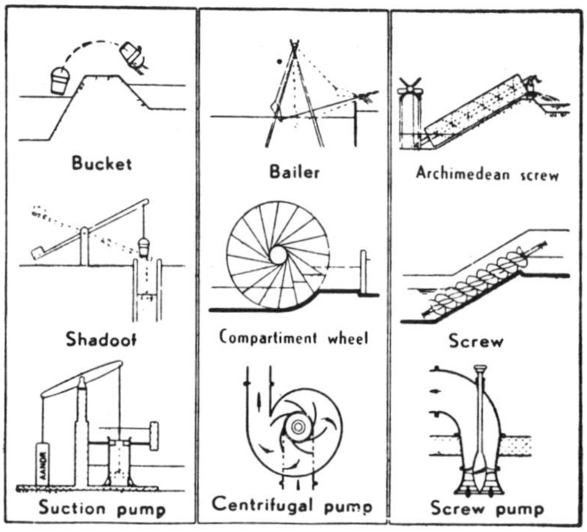

Fig. 7.
The evolution of water-raising machinery
(After Dibbits, Nederland-Waterland fig. 105)

figs and vine. Therefore irrigation was applied only if the climate was all too arid or if the crops demanded large quantities of water. It supplemented "dry farming" locally for instance to grow onions, cabbages, lettuces and pears. In certain cases vineyards were also irrigated. "The harsher vines need to be watered, at all events in the Fabii district of the territory of Sulmo in Italy where they irrigate even the ploughland; and it is a remarkable fact that in that part of the country water kills herbaceous plants but nourishes corn, and irrigation takes the place of a hoe for weeding," PLINY tells us (131).

Irrigation -water was supplied from aqueducts, generally speaking. We have an inscription with a plan of an aqueduct (the Aqua Crabra near Tusculum) which gives the names of the properties irrigated, the number of pipes supplied and the hours they could be opened.

In North Africa, where the climate and conditions were more like those in the Ancient Near East irrigation was much more common (132). Springs were used to irrigate small hill-plots in Numida, the schedule points to small properties (133). Everywhere in Algeria we find vestiges of Roman hydraulic works. Streams were dammed, water stored in large cisterns, ponds and underground tanks and finally distributed by canals and aqueducts. In the home country irrigation usually failed as the agricultural technique was not the correct one. Only when the Arabs introduced such crop as rice in the Western Mediterranean did irrigation techniques penetrate Italy.

In the ancient Greek world the cultivated area was restricted and potentially fertile land was covered with shallow lakes, swamps and marches. Attempts to drain these were made from the earliest times particularly in the Copais district of Boeotia (134). Lake Copais was in Antiquity a vast reed swamp. In November the water-level rose to about 2 m above the present surface. The natural outlets of this depression were a number of swallets (Greek katabothrai, Latin emissari), natural subterraneous tunnels and crevices in the surrounding limestone hills. These discharges were subject to blocking by the frequent earthquakes and the ancients have often attempted to make new tunnels. A repetition of high water levels during two or three years spelt evil according to ancient tradition (135).

The Minyan Orchomenians are said to have drained the basin (136) but Herakles was said to have blocked the emissaries (137). In fact the region was often compared with Egyptian Thebes. It is now generally believed that in prehistoric late Helladic Times (1400 B.C.) the waters of the rivers falling into the lake were intercepted at the very edge of the basin and conducted in embanked canals to the emissaries. Often the rocky ledge was used as one bank of the canals. These canals fell into disrepair and as late as A.D. 40 a certain "Epaminondas" paid 6000 denari for "the cement used to repair the dyke" (138).

Alexander the Great's "mining engineer" Crates of Chalcis, though the subterranean channels filled up again, ceased clearing away the obstructions because party strife among the Boeotians, although", as he himself says in the letter to Alexander, "many places had already been drained". (139). Crates is also called a "taphrorychos", canal-builder, for what he actually seems to have attempted was constructing a canal to convey the flood-water of the Hercyna into lake Copais without overflowing the cultivated land in the south, but none of the texts say anything about a tunnel attempted by Crates.

Empedocles (450 B.C.) is said to have drained the swamps near Selinus and to have improved the climate and harvests of Agrigentum by blocking a windy rift in the hills (140). We also have the inscription (141) of a contract of the city of Eretria (Euboea) and a group of private capitalists headed by Chaerephanes (c. 320 B.C.), which latter promised to recover fertile lands by his hydraulic works which in case of success would be leased to him for ten years on payment of ten talents.

Fig. 8.
Compartiment wheel used by the Romans to drain the copper mines of Southern Spain.

The Romans were far more active in the field of drainage (142). Many Italian rivers like the Arno and the Tiber can hardly contain the torrential spring floods and form swamps near their mouths or regularly flood the fields along their banks. At present 1.790.000 ha of land in Italy qualify for drainage, of which 1.000.000 ha have not yet been drained. Hence in Rome itself the Tiber was regulated towards the end of the third century by building a stone embankment in three stages, stepping back at each stage, which made the width of the waterway 62.5, 97.5 and 135 m. It proved a decided succes over the modern straight quay and a similar construction was advocated by Leonardo da Vinci for the regulation of the Arno to win through after a more "modern" solution had failed to satisfy the city authorities who at first sight had rejected Leonardo's plans.

The drainage of several districts around the highland lakes or below

Velitrae seem to date back to the early republic. They were designed to improve farm lands over large areas and testify the presence of great landlords who could command large resources of labour. This drainage was achieved by water-tunnels (emissaria, cuniculi) often subterraneous to tunnels of the qanat-type with air (and construction) shafts at regular intervals. They occur in the Pomptine Marshes south of Rome and at several places in Etruria.

Lake Albano, in the crater of an extinct volcano, was provided with an outlet tunnel soon after the taking of Veii (396 B.C.) by the construction of an 1800 m tunnel with airshafts every 35 m, its area being 2 by 1.10 m. A 15-mile open channel then conducted the water to the Tiber. A century later Lake Velinus and the plain of Raete was drained by an 800 m canal partly cut in the rocks. The drainage of the large Lake Fucinus was undertaken by the emperor Claudius in 41 A.D. and completed by 30.000 men in eleven years (143). The water of this depression in the mountains, the level of which varied greatly, was carried through Mt. Salviano to the Liris (the present Garigliano). This 5640 m tunnel, with airshafts at regular intervals, had an area of 16 square yards. It cost 400.000.000 sesterces and added 50.000 acres to the Imperial estates.

There were several attempts at draining the Pomptine swamps south of Rome, but no definite results were obtained and in later Antiquity the population of the district were decimated by malaria until properly drained during this century according to plans submitted to Pope Urbanus VII by a Dutchman van den Houten and partly executed by two other Dutch engineers van der Pollen and Meyer some threehundred years ago!

The Senate entrusted the consul Ceteghus with this task (160 B.C.) but apparently with little success (144). Julius Caesar took up these plans, but again nothing came of his ideas of draining them by digging a canal to the Tiber (145). Augustus had a canal, the "Cavo di Augusto", made which has now disappeared. Nero had the intention to build a 160 mile ship-canal from Ostia to Lake Avernus, in which two five-bank ships could have passed each other (146). A tunnel on the north side of lake Avernus still testifies of the work of the famous engineers Severus and Celer, though the canal was never finished beyond Misenum. There was an older canal from Terracina to Forum Apii (147) in which Romans sometimes took pleasure-rides (148). Theoderic the Great again ordered Decius (500 A.D.) to drain these marshes, but this attempt failed and the region was abandoned by the population.

The drainage of the Po valley was much more successful. This project formed part of the policy of settling ex-soldiers in the new colonies when northern Italy was conquered. Different consuls partook in the gradual drainage of the Po valley. M. Aemilius Scaurus "drained the plains by running navigable canals from the Padus (Po) as far as Parma (149)", in 109 B.C. before constructing the Via Aemilia connecting this district with Rome. Thus good farmland was obtained in "Cispadane Gaul" (the present Emilia) north of Parma and Modena. Marcus Aemilius Lepidus then drained the district round Bologna and Piacenza and Aulus Postumus the neighbourhood of Cremona. In 31 B.C. Octavius canalised the Adige between Ferrara and Padua of which work we have two inscriptions (150). One of these reads: "Gang of Quintus Arruntius Sura: Foremen are Quintus Arruntius and Gaius Sabellus: the bail is deposited with Titus Arrius: the total length is 4214′." The width and height of the dyke is not given. The other inscription specifies that 88 men each building 27.5′ of dyke are to build a stretch of 2.5 km of properly protected river bank.

Part of these works was also protecting the V. Aemilia by a dike against the encroachment of the river Reno near Bologna (151). The emperor Claudius finally crowned the work by drainage north of the Po and by the building of the Via Altinata (152). These projects also included the drainage of the marshes of Ravenna which town became increasingly important when the Adriatic fleet was stationed there (153).

Nor were Roman drainage and river regulation projects confined to Italy. The army canalised one of the mouths of the Rhône (Fossa Marina) in order to safeguard the regular supply of the army from overseas (154). Though this canal silted up in the first century A.D. Arles remained an important army centre with good communications for inland shipping. In the fourth century A.D. there is a general complaint about the neglect of drainage in the lower parts of Gaul (155).

In Britain the Romans attempted to reclaim land in the Fen district on a larger scale. They built an embankment to protect the flat lands of the Wash from the sea and other dikes enclosing flat marshy land at the mouth of the Usk. They built an 8 mile canal, "Car Dyke" connecting the Cam below Cambridge with the Ouse halfway between St. Ives and Ely. Another "Car Dyke" was a 25 mile canal (60′ wide) from Peterborough on the Nene to the Witham 3 miles below Lincoln. The Fosse Dyke connected the Witham and the Trent. In connection with these drainage canals a 60′ causeway was built along some 30 miles of their banks.

In Holland Drusus in 12 A.D. (156) possibly laid a mole in the Rhine west of Cleve to keep excess water from flowing through the Waal and flooding the southern part of the Betuwe district (between the rivers Rhine and Waal). He then built a dyke to protect the eastern flank of the Betuwe and dug a canal between the rivers Rhine and Yssel to relieve the Rhine of surplus water (157). However, this problem is still vividly discussed. About 45 A.D. Corbulo connected Rhine and Meuse with a 23 mile canal running along the later site of the Hague (158), providing his ships a safe passage avoiding the sea.

Irrigation and drainage in the middle ages

The Arabs should be credited with important developments of irrigation in the Western Mediterranean. Within a short period after the death of Mohammed the Mediterranean threatened to become virtually a Moslem sea. As they pushed westwards the Arabs introduced the irrigation methods and machinery of the Ancient Near East together with certain crops which could not have been grown with the typically classical agricultural methods. The Romans had imported rice but had never grown it on a large scale. Now the Arabs started to grow it on irrigated fields in Sicily and Spain, whence it came to the Pisan plain (1468) and Lombardy (1475). They grew the bitter orange in Sicily in 1002, the sweet orange came to Spain and Italy in the fourteenth century. Cotton and sugar-cane also came west with the Arabs.

Water-raising machinery was of course known to the Roman agronomists but it never played a large part in their economy. Still the shadoof was used by the Spanish gardeners of the sixth century A.D., it was known as the "ciconia". Slowly it spread north where we find it in Germany and Hungary in the fourteenth century and on some pictures by Breughel.

The Arabs greatly extended the irrigated area in Spain and Sicily. The "Moors" of Spain knew perfectly how to drain rivers and how to irrigate their fields by systems of branch channels with an efficient distribution of the available water. The big water-wheels at Toledo may date back to their time. This heritage was taken over by the Christian conquerors.

The irrigated area in the Ebro valley was materially increased in the thirteenth century, after the conquest. The Spanish "huertas" are perfect examples of the ancient small irrigation areas. In Rousillon (southern France) irrigation methods and crops were introduced in the eleventh and twelfth centuries. A century later we find a system of canals, water-wheels and water-mills for irrigating crops. In Italy irrigation works appear in Lombardy in the twelfth century and some hundred years later in Emilia. The irrigated meadows ("marcite") round Milan date of the fourteenth century. Here the Cistercensians introduced the use of city refuse and sewer-water as fertilizers about 1150 A.D.

The Arabs were also responsible for the spread of the noria and other forms of water-wheels. Sometimes the West even calls them by the Persian name of "doulab". The early western terms for this machinery are a particularly apt illustration of classical tradition being overrun

Fig. 9.
Al Jazari's picture of a muledriven wheel of pots

by Egyptian, Arab, Berber and Persian elements which spread west and then north by the way of Sicily and Spain.

The farmlands of France were extended by serious attacks on the forests of Normandy and Maine which began in the tenth century but the next two centuries until the death of Saint Louis form the real age of reclamation which about 1300 still goes on in the south-west of France. However, the gradual deforestation of Europe was not the most important front. The main attack was directed against marshes, rivers and the sea and here drainage was of course the dominant factor.

In England the meadows extended on the marshes in the Middle Ages but this was piecemeal encroachment on the fens not the consequence of organised engineering projects. The drains and gates of the Lincolnshire fen region continued the Roman frontal attacks on the sea. They gave rise to artificial divisions, the inhabitants of which, "inlikes", all had to contribute to the upkeep of the ditches, drains and sea-walls.

The main developments of this type, however, took place in Flanders and the Low Countries on the mouths of Scheldt, Meuse and Rhine whence these techniques spread eastwards in the twelfth century first to the lowlands west of the Elbe and then east of this river. The evolution of drainage techniques can best be studied in Flanders and the four sea-provinces of the Netherlands (Zeeland, Holland (originally the name of a district near the town of Dordrecht), Friesland and Groningen (159)). The situation in Roman times is well illustrated by this passage from PLINY, who visited this region of north-western Europe (160): "In the land of the Chaucians twice each period of a day and a night the ocean with its vast tide sweeps in a flood over a measureless expanse, covering up Nature's age-long controversy and the region disputed as belonging whether to the land or to the sea. There this miserable race occupy elevated patches of ground or platforms built up by hand above the level of the highest tide experienced, living in huts erected on the sites so chosen and ressembling sailors in ships when the water covers the surrounding land... They scoop up mud in their hands and dry it by the wind more than by sunshine, and with earth (peat, turves) as fuel warm their food and so their own bodies frozen by the north wind."

However, this is only part of the picture for in several regions of the Low Countries the sea did not encroach on the land this much during the fairly dry sub-boreal period (1500 B.C.—500 A.D.). Many of the mounds mentioned by PLINY grew up from the refuse collected in inhabitated sites. When, however, the climate grew moister from 500 A.D. onwards and the land sank some 10 cm per century, the situation grew far worse. The mounds were now artificially raised to form safe sites in case of storms and flooding of the low meadows. These mounds are called "terpen' in Friesland and "wierden" in Groningen. They are characteristic of these northern regions where we find no less than 1260 mounds (varying in area from 5 tot 40 acres) on a stretch of 60 by 12 miles. Their memory is enshrined in the names of such towns as Leeu*warden* and Bols*ward*. As the "water-wolf" was more often on the hunt encroaching on the clay and peat only partly defended by dunes, men could no longer rely on the defensive only. For even in the south

(Flanders and Zeeland) floods became more frequent. The frontal attack on the sea was at first unorganised. Dikes were built on the danger spots and because the bays and creeks were still small and floods relatively low, lower dikes were still sufficient. This building of dikes started as early as the seventh century and by 1000 A.D. Friesland was relatively safeguarded by dikes. This improvement is shown by the fact that the mounds were no longer artificially raised after about 850 A.D. In his *Life of Boniface* WILLIBALD tells us about the building of a new mound at Dokkum for the church and presbytery during Pepin's reign (c. 700 A.D.). After the storms of 839 some more mounds were raised, but this seems the last notice of the kind in our annals. The offensive had now started.

The *Notitia vel commemoratio de illa ewa quae se ad Amorem habet* written about 800 A.D. mentions as article 38: "Si quis sclusam dimiserit quando suus comes ei commendat..." which shows that dykebuilding belonged to the common duties in those days.

In Flanders and Zeeland land was gradually wrested from the sea. The saltings and muddy flats on the sea-shore were first naturally consolidated by seaweed and then subsequently glasswort and sea-startwort gradually raised the sea-side meadows ("schorren") on which finally the peculiar salting-grass (Kweldergrass) came to grow. These "schorren" suffered less and less from periodical tidal floodings and they could be used as grazing-grounds for sheep. In this region we find typical man-made mounds, some 10—12 m high with steep slopes, the "vliedbergen" or "hillen" which ressemble the Norman moated mounds. They served as refuges in case of sudden storms. Archaeology has proved that they were built in the fifth to seventh century A.D. Towards the end of this period men began to endike the "schorren" and they were no longer needed.

Thus land thus acquired was first used as grazing grounds for sheep. Then it was gradually drained properly and either ploughed or used for horses and cows. The lush grass growing there was ideal for the breeding of strong heavy breeds. This reclamation which certainly started very early became a strong factor in the policy of the counts of Flanders, who from the twelfth century granted to abbeys and chapters in the maritime Flanders and Ypres region new lands which were still exposed to the sea apart from wasteland in the heath and woodland zone of ypres and Bruges.

Here in the south or in the rest of Europe this landreclamation was started and led by religious houses. Thus the Benedictines worked on the irrigation of Roussillon and in Saintonge, Maine, Ile de France and Bavaria. More to the north there were the Praemonstatencians and above all the Cistercians who cultivated the waste-land and left the traces thereof in such place-names ending in "-sart" or "-rode". Their main works are found in Germany between Rhine and Elbe, in Saxony, Thuringia and finally in Lusatia and Bavaria. The Cistercians worked with lay brethern or "conversi". However, several reclamation jobs in Flanders were ordered and directed by a "locator" of knightly family or a "clearance contractor".

By the tenth century the ancestors of the later "dike-masters" had gained considerable experience and "like the Flemings between Wit-zand and Bruges fearing the flood that rises on them raised a bulwark to hold the sea" (161). Their skill became famous beyond their home-land and as early as 1106 the Archbishop Frederick of Bremen calls in colonists from Holland to reclaim bogs north-east of his see. It should, however, be realised that upto 1500 more land was lost than gained and held, and this is particularly true for the earlier part of this period. As the land was endiked the storms grew heavier and the sea, induced by the sinking of the land, encroached more and more upon the old land. The lake Flevo of Roman date became the "Almere" and finally reached its largest extent as "Zuyderzee" about 1300. During the storm of December 14th, 1287 over 50.000 people were drowned between the town of Stavoren (Friesland) and the Eems, and it was several years "before the people ventured on the land again". In 1404 heavy storms formed the Braakman bay in northern Flanders. In Zeeland the "land of Reimerswaal" was forever lost to the sea. Near the mouth of the Eems the Dollard bay was formed. In 1270 the men of Dordrecht had ventured to dam the mouth of the Meuse, endyking it for 30 miles and leading its waters northwards into the Rhine. But the "water-wolf" wrought his vengeance in the night of November 18th, 1421, when the St. Elizabeth Flood broke the dykes of the Hollandsche Waard drown-ing 65 villages and 10.000 people, creating the present Biesbosch ("Forest of Rushes") a region still to be reclaimed.

This selection of calamities that beset the early dike-builders will show how merciless the battle was and how slowly the attack with the iron spade alone proceeded. Only when the art of dikebuilding was thoroughly mastered and reclamation organised as a function of society did gains become more permanent. "He whom the waters harm will throw them back" says the Dutch proverb. Though the *Lex Frisionum* of 802 does not yet speak of sea-walls, so much the more do later codifications of ancient law, when a man was excused from attending the Thing "if water and wind had turned against him so that he should attend to the dike". By gradually encircling the threatened land by dikes man made the "polders", units which could discharge their water into the sea at ebb-tide through gates of sluices. The word "polder" is derived from the Frisian "polle", meaning something that is raised above a certain plane; it began to be used about 1400

The main law of this drainage was "shorten the coast and close the bays, creeks and gaps". This was indeed the general programme of the Low Countries. The Middelzee which cleft Friesland in two was endyked during the thirteenth and fourteenth centuries by the monasteries of Anjum and Mariëngaarde, the old sea-dikes now became inner-dikes. About 1300 the great dike encirching West-Friesland (prov. Holland) was ready, it is mentioned in a document of the bishop of Zuden and other counselers of the Court of Holland (1319). Early in the thirteenth century certain bays in the north of Holland had already been dammed (the Zaan near Zaandam, the Leek near Medemblik). In 1315 the Beemster was dammed at Schardam, the Krommenye at Nieuwerdam (1357), the Purmer Ee at Monnikendam (1401). By the end of that century most of southern Holland was endyked, and land was gained in the same way in Zeeland. From 1200—1500 over 285.000 acres were gained and held.

But the sea was not the only enemy. With the rise of the cities peat was dug in the low country behind the dunes. The shallow lakes thus formed fanned by the storms encroached on the land and finally covered a great part of the land. These lakes had to be endiked, drained and recovered as farmland. The retaining dams became dikes of a new type of polders. In this region the villages are built on the dikes or beyond the ancient lakes. In the polders reclaimed after 1500 which were lake-bottoms originally, the villages lie in the polder itself.

Upto the fifteenth century salt was made by incinerating peat. This "saltburning" ("selbarnen") (162) of peat gave ashes from which salt was leached. It was common custom long before the Carolingian period and some of the towns in Zeeland like Zierikzee grew rich on this salt trade. We find it too in Enkhuyzen and Schleswig. Hills of extracted ashes were later used in glass manufacture. This custom of "darink-delven" (derrie, darink = peat) was restricted during the fifteenth century and finally forbidden by Charles the Fifth in 1515, when supplies of salt from Spain had long been plentiful. This cutting of peat destroyed good land and often endangered the polders by inducing a higher water-table. Again the cutting and destruction of inland former sea-dikes often endangered the land. This was finally forbidden except with the express consent of the proper authorities (1595). Finally the shifting of currents and especially very low ebb-tides did often endanger dikes that were none too solidly constructed.

The dykes, originally private ventures, became planned and organised attempts at keeping the "water-wolf" out. This is clear from the Rüstinger Rechtsregeln, a twelfth century codification of the "ewa" or Law of Eternal Rights for Friesland which say: "This is also the Right of the Land that we Frisians have to found a Sea-Burgh and to maintain a Golden Ring (Hoop) all around Friesland in which each yard of dike is equal to the other where the salt sea swells both day and night. Also we Frisians shall defend our country with three weapons: with the spade, with the handbarrow and with the fork: and we shall defend our country with the sword and the spear and the brown shield against the high helmet and the red shield and against every unjust lordship. Thus we Frisians shall hold our country from end to end, so help us God and Saint Peter."

Soon it became clear that the construction and maintenance of the dikes and polders was a task of the community and that each individual should play his part in this scheme. This task "rested on the soil", hence farmland was "infected with dike". Should the farmer be unable to fulfill his share in this maintenance he could "stick his spade in the dike" and whosoever drew this spade out took over his task but also acquired his farm. Gradually more and more polders sprang up and cooperated in larger units when they had common interests in draining and discharging their water either through a common gate into the sea or in stages by first discharging the polderwater into a dammed bay or stream, called "boezem", and thence into the sea. This cooperation of polder with joint interest was consolidated by the formation of "hoogheemraadschappen (waterschappen)", managed by the "dijkgraaf" (often appointed by count or bishop) and "heemraden (gezworenen)". They regularly inspected ("ruwaarschouw") the dikes and gates, and allotted the portions ("verhoefslaging, verstoeling, verkaveling") for each farm to maintain either with labour or money.

By the thirteenth century the hoogheemraadschappen such as Rijnland, Delfland and Schieland were already formed. For the next 250 years they consolidated their position against the landed gentry and the towns, who tried to dodge their responsibilities. They had the support of the central authorities, who once the battle was won, centralized and established their authority over these drainage units during the period of 1450—1568. By then man's attack on the sea had gained new impetus for town councils clamoured for charters to build dikes and reclaim land. Now that all the *land* had practically been safeguarded, the inland *waters* formed by peat-delving activities became more potent dangers and the rich merchants formed private companies to reclaim these lakes and exploit the farmlands won from the water. A more systematic planning of the reclamation of the lakes of northern Holland was started by Lamoraal, count of Egmont (1522—1568) during the reign of Charles the Fifth (1516—1556), when the first lake the Dergmeer (42 ha) near Oud-Karspel was drained. This conquest was possible because of the accumulated knowledge of dikeconstruction and the introduction of the windmill-driven scoopwheels to pump the rain- and seepage-water of these deep polders. The period of well-centralised drainage policy had begun.

Dikes, windmills and sluices

Dike-construction in the Middle Ages was the result of practical experience (163), which was not codified until many centuries after the first dikes were built. Actually the oldest handbook on these matters written by Andries Vierlingh in 1570 was never published until some thirty years ago (164).

The oldest *clay-* (earth-) *dike* was simply a mass of lumps of boulder clay consolidated by treading oxen or horses. Even the Rüstinger Rechtsregeln of the twelfth century does not mention handramming. The subsoil ("dijkzate") was usually carefully denuded of any traces of vegetation and in front of the dyke there was usually a stretch of foreshore. Some dikes (mostly in Zeeland) had an earth core plastered with clay on the slopes ("slikkerdijk, slijckdijk"). Vierlingh reports that certain dykes had stepped slopes as they were built up and that finally the 1.5:1 slope was plastered with clay and planted with grass. In 1328 Henry of Brederode allowed the people of Callantsoog to graze their cows on the dikes and this is still common practice. Later the "Krammat" was adopted in Zeeland to protect the dikes. Straw was piled several cm high perpendicular to the axis of the dike and fastened to the slope by bundles of straw ("Krammen"). Bundles of osiers ("rijsbeslag") was also applied in the same manner. This form of protection gradually took the aspect of the modern mat of osier-bundles weighted with blocks of basalt.

The slope facing the sea bore the brunt of the storm and especially in the north where the slopes were steeper the clay-dike needed better protection. This was provided by the fourteenth century *seaweed-dike* ("wierdijk"). This was a clay-dike with a frontage of sea-weed to break and absorb the action of the waves. The bundles of seaweed were piled step-wise against the vertical front of the dike and even 1′—2′ higher than the top. By its own weight the sea-weed was compacted and by its heating gradually formed a compact mass. An order of the Count of Holland (1518) specifies that the sea-weed frontage should measure 2.40 m at the bottom and 1.80 m at the top. As early as 1319 the bishop of Zuden ordered the Frisians to repair their dikes: "with clay the clay-dike and with seaweed the seaweed-dike". The cost was fairly high, 200 guilders a meter and it gave difficulties with turbulent waters or ice-drifts, but generally speaking it was a decided improvement on the clay-dike and it remained in use until well in the eighteenth century.

In Zeeland and other places where sea-weed was not available reed was used instead. The *reed-dike* had a frontage of polls of reeds piled up with the roots facing the sea. The strength of this protective layer was determined by the length of the polls, there was no compaction and the reed was liable to rot over the water surface. Every five or six years it had to be renewed. In 1402 Albrecht of Bavaria mentions such a "dike between Winkel and Werfershove (south of Medemblik) made of reeds and sods" and its use along the southern coast of the Zuiderzee continued until about 1750.

During the fifteenth century more lasting forms of protecting the dike were introduced. First of all there was the *pallisade* ("paaldijk, platinge") which was often erected in front of the seaweed-dike. Philips of Burgundy ordered the dike from Amsterdam to Muiden to be protected in this way (1440) by a "bolwerck met platinge", in 1466 he decreed that the sea-dike west of Medemblik should be protected over a length of 4.5 km as quickly as possible "with piles and iron like a bulwark as far as it will prove necessary and economical". Somewhat earlier in the fifteenth century rows of piles were used as breakwaters in Friesland. The piles were generally 7 to 10 m long and 30 cm in diameter. They were held together by double girders and iron bolts.

Right at the end of our period a new form of dikeprotection came into use called "Krebbingen" which consisted of two rows of short piles, a few feet apart, which space was then filled with faggots or fascines held down by stones. This was the ancestor of the later fascine work. In the fifteenth and sixteenth centuries a few primitive attempts were made at "harnessing the dikes with stones" and VIERLINGH reports that the best way to protect the frontage of the dike was to lay the stones like the tiles of a roof but that this excellent protection was too costly for general use. During this period the planting of the dunes with beachgrass was introduced to prevent sand-drift.

Once the attempts to keep the sea out had succeeded, a new task began. Rain-water and seepage water had to be drained at regular intervals. In early times the canals collecting the surplus water from the ditches discharged this water to the sea at ebb-tide or in stages, first into the endiked bay or canalsystem ("boezem") and thence into the sea. This was effected by sluice-gates or doors or by weirs built into canals or dikes. In the days of Gregory of Tours (c. 580) and long after a "sclusa" was a weir and not a sluice. Such weirs in canals were of course a hindrance to shipping and ships had to be towed over these weirs with windlasses or by man-power. Often one had to resort to transshipping. These lifts for ships often took the form of inclined planes ("overtoom") over which the ships were hauled, which are illustrated by Zonca and other Italian engineers about 1600 and which still exist in some parts of the Netherlands (165). In the case of polders the excess water was often discharged through a tube or conduit through the dike which could be closed by a movable watertight door moved up and down with chains. Such doors which first moved between wooden posts encased in the dike were later held in masonry constructions. Such well-built flood-doors appear in the early Middle Ages (1065!) and they are discussed in detail by VIERLINGH (164).

The inclined plane or sluice gate was a definite improvement on the old situation in which the polder-dike formed a serious obstruction for shipping. As the canal-system became increasingly important to trade in a region where 85% of the goods are still transported by water a more efficient solution was wanted. The sluices divided the canals into stretches of water of the same level and the ascent or descent from one pool into another can only be achieved by the lift-lock. This is a complex of two gates with a chamber in between which can contain at least one vessel to be transferred from pool to pool. Such locks also permit rivers of irregular slope to be converted into series of horizontal levels with vertical steps overcome by the locks.

The story of the pound-locks in the Low Countries is complicated by the fact that so few technical details are known of the medieval examples mentioned in our documents. We know that the first set of wooden sluice-gates was built by the city of Bruges in the Zwin near

Damme about 1234 by Dutchmen (165). It was rebuilt in stone as a pound-lock in 1394—1396 and the documents speak of the sea-gate ("soute deure") and the fresh water-gate ("varssche deure"). The next example is mentioned in an edict of Count William II of Holland of April 5, 1253. It mentions a plan to build a sluice ("spoije") in the dam at Spaarndam near Haarlem which should be 24′ wide. The plan was executed about 1285 and provided for "the easy passage of sea-going (Zuyderzee) ships." Count William III in 1315 fixes the tolls to be

Fig. 10.
A set of gates along a river. (Codice Laurenziano, Florence)

paid by ships passing the locks (167). The contention that there was a pound-lock in Amsterdam in 1220 is unconfirmed by later investigators (168). In Vierlingh's days the locks of Spaarndam were again rebuilt. All these locks had vertical gates which were drawn up and lowered. Vierlingh mentions that polders upto 400 ha should have a 8′—9′ discharge-gate and larger ones upto 800 ha, double gates each of this size. Another set of gates was built at Vreeswijk (province of Utrecht) but it did not serve as a pound lock for ships until 1433. About that time we hear of other pound locks ("kolk", "schutting") in the Netherlands.

The evolution of the pound lock in Italy is much better known (169). In the thirteenth century the canals of Milan have "conches" which were single barriers or weirs to make them navigable. Their flash locks which were single gates like the flood gates and sluices of the Low Countries and France, were opened at certain times and the boats plunged downwards or were drawn against the current by men or windlasses. This tedious operation was used in 1395 to transport marble

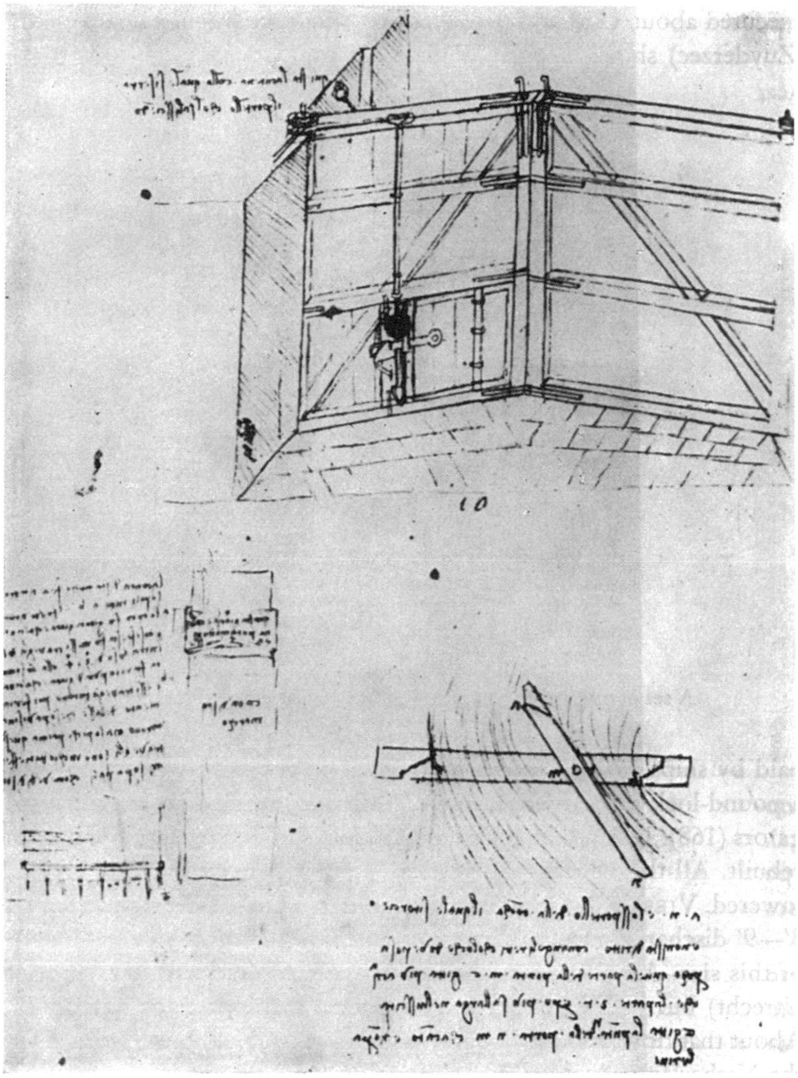

Fig. 11.
Leonardo's design of the lock at San Marco, as he rebuilt it (Codex Atl. 240r-c)

down the Naviglio Grande for the building of the Milan cathedral. The one gate then in the canal was used to level its waters and those of the interior canals (a difference of several feet). Once this was achieved the gate was opened and the boats could pass. During the operation it was strictly forbidden to draw water for irrigation purposes. In 1445 we hear that "a lower lock was recently constructed" and thus the operation was made far easier if the pool between these two flash-locks was used as a large lock-chamber. Apart from this there was a second lock on the Naviglio Grande before 1445.

The experience naturally led to the suggestion to divide the rivers and canals with too large a slope into pools by means of such flash locks. This step is suggested in the manuscript entitled *Trattato dei pondi, levi e tirari* (170) which was written about 1459—1463 by a practical engineer and which has marginal notes in the hand of Leonardo da Vinci. However, we do know of such application of vertical gates with windlasses as given here near Mantua in the twelfth century and in Flanders, we have the two sets of gates on the Naviglio Grande and this text, which says: "Let us suppose that the first part of the river has a drop of 30 piede; construct at that point a high door in the manner of a portcullis, with windlasses to raise, it, and in this manner lay off the entire length of the river and all its falls with such doors." The further description of the handling of boats shows that each pool is treated as a lock-chamber. Such single gates at intervals along a river had been in use for quite some time before 1459.

Though this would form the logical stage preceeding the actual invention of the pound-lock, the latter does fall earlier. Though we can not obtain absolute certainty on this point, it would seem to have been made by LEON BATTISTA ALBERTI (1404—1472) in his *De re aedificatoria* written in 1452 and printed in 1455. After describing the methods of drawing a boat against a water fall he continues: "Also, if you wish you can make two gates cutting the river in two places at such a distance one from the other that a boat can lie for its full length between the two; and if the said boat desires to ascend when it arrives at the place, close the lower barrier and open the upper one, and conversely, when it is descending close the upper and open the lower one. Thus the said boat shall have enough water to float it easily to the main canal, because the closing of the upper gate restrains the water from pushing it to violently, with fear of grounding."

Before the above-mentioned texts were found many conflicting theories have been presented on the evolution of the lift-lock, even in Italy. Now we know that neither Philip Maria Visconti the Venetian (1440) nor Dionizio and Pietro Domenico of Viterbo (1481) were the first to build a pound-lock in Italy. Bertola da Novate was the first canal engineer of note who, commissioned by Francesco Sforza of Milan, designed and constructed the new Bereguardo canal with at least 12 locks (1452—1458), the Martesana canal from the Adda river to Milan (1457) with two locks and three more canals with locks, though he was certainly not the inventor of the lock.

Leonardo da Vinci, who is said to have completed 6 locks at Milan (1497), was the inventor of the mitered gates and the wicket. His note-books illustrate the vertical, horizontal and mitered doors together with many details on the construction of sluice-gates and lock-doors and the proper mechanism for opening closing and installing them (171). He is in fact the creator of the modern ship lock later discussed by ZONCA in Italy (172) and by STEVIN (173) in the Low Countries. We have drawings of French and Flemish basins with sets of sluices of the sixteenth century (174).

Sluices and locks were efficient as long as the polders represented land won back from the sea. When, however, lower grounds and the lakes in the ancient polders were drained, discharge of water by natural flow at ebb-tide was no longer possible and also the relative amounts of seepage water grew considerably. Hence pumping had to be resorted to. The low new polders usually threw their water ("uitslaan") into a canal built round the polder dike ("ringvaart") which then discharged its waters in the usual way. Sometimes the new polders were so deep that the pumping was achieved in stages ("maaltrappen") by a series of windmills ("molengang") each of which stepped up the discharge water until it finally reached the discharge canal or bay ("tussenboezem").

Until steam-engines were introduced to drain polders (1787, polder Blijdorp near Rotterdam) the wind-mills drove the pumping machinery. The windmill reached the Low Countries as a flour-mill (175) and was not used to drive pumping machinery or scoopwheels before the fourteenth century, which is actually the period when this became necessary for the reasons given above. A document of the reign of Count William IV (1344) mentions the sale of cornmills and wind-driven water-mills (marsh-mills) in Drechterland, without, however, giving details on their construction. Soon we hear about more marsh-mills near Brielle (1394), Schieland (1434), Delfland (1440), Enkhuyzen (1452) and Arkel (1456). In 1408 the secretary of the treasury of Holland mentions a trip of drainage officials ("hoogheemraden") from Delft to

Alkmaar to inspect the new marsh-mills built by Floris van Alkemade and Jan Grietenzoon "to throw water". A tax-document *Informatie op de verponding van het jaar* 1494 informs us that Holland had several such marshmills by that time, but even in 1514 there were but 8 north of Amsterdam, where the drainage of the lakes had not yet begun.

The old post-mill built to grind corn had to be recast hence the new typical marsh-mill ("wipmolen, wipwatermolen"). The earliest marsh-mills were probably tower mills such as the octagonal "Old Mill" built in the polder Oud-Beijerswaard in 1438. Between 1438 and 1453 a "New Mill" of the marsh-mill type was added to the drainage machinery of this region south east of Ridderkerk. The State Archives have yielded a decree of May 13, 1430 in which the district called Krimpenerwaard was allowed to make plans to put their drainage problems in order. This was no easy task for as late as 1550 the Emperor Charles V commanded the bailiwick of Schoonhoven to remove such high trees as imperilled the proper functioning of the marsh-mills. It would seem that the mills in this district were built between 1430 and 1450. The mill of Bonrepas formerly believed to have been built in 1430 was erected in 1449 only but it served the region efficiently for well over a century. This mill proved so efficient that by the end of that century over 17 marsh-mills pumped their water into the Vlist.

This "wipmolen" type of marsh-mill drove a scoopwheel in a simple way. The "brake wheel ("kamwiel") on the "windshaft" ("molenas") engages with a "trundle, wallower" ("bovenbonkelaar") on top of the upright shaft ("koningsspil") which with its "spur wheel" ("benedenbonkelaar") at its lower end drove the bevel pinion ("onderwiel") which turned the "scoop wheel" shaft by means of a "pit wheel".

The scoop wheel, descendent of the scoop and ancestor of the centrifugal pump, pushes the water up and flings it away. The 6 m wheel revolves about 5—7 times a minute between two walls ("krimpmuren") and drives the water upwards against a race of circular shape ("opleider") until the threshold of the discharge canal is reached, the entrance of which can be closed by a gate ("wachtdeur"). The floats are immersed ("insteek") 60—90 cm in the polder water and can push up the water about 2 m. The velocity of the floats is about 2.0—2.5 m-sec. at the end of the floats but the miller measures it by counting the number of sails ("enden") that pass a certain point per minute. In general marsh-mills will discharge about 10 m^3 per minute with a low wind, 25 m^3 with a good wind and 40 m^3 in a strong wind (e.g. about "60 enden").

The marsh-mill turned on a post and the windshaft and sails on one side and the tailpole and stairs on the other balanced the mill. They were set to face the wind. It is not known whether horsedriven scoop-wheels preceded this marsh-mill, as no documents mention them though some authors do. There were, however, a few horse-driven Archimedean screws until here too wind-power was used. In Antiquity the cylinder of the Archimedean screw was moved. This form called "ton-molen" was known in the Low Countries but it was not geared to marsh-mills. These mills did, however, drive Archimedean screws, the screw of which revolved ("vijzel"). This ancestor of the screwpump was geared to the marsh mill in the fifteenth century too. In the days of STEVIN the choice had already fallen on the "vijzel" ("gootwentel") rather than on the "tonmolen" ("tonnewentel"). The screw is usually placed under an angle of 30° (upto 38°) with the water, its diameter is 1.50—1.80 m. With the usual 40—50 revolutions a minute it throws water 4—5 m high and a "vijzel" of 1.65 m diameter will pump 40 m^3 per minute. Further forms of pumping machinery date of the sixteenth century. In the later Middle Ages the scoopwheeled "wipmolen" was undoubtedly the most popular marsh-mill. Smaller forms called "staart-molen" date of the earlier part of the century, the full-blown size coming into its own after 1430.

As drainage techniques grew there was also a growing need to improve waterways and canals by dredging. Though harbours and rivers near Rostock seem to have been dredged as early as 1288 (176), we have the first description of a sailing dredger at Middelburg (Zeeland) called "the Mole" ("Krabbelaar"). The efficient dredging machines are of later date. Dikes, locks, marsh-mills and dredgers were now available to control the water-level in western Europe and to regulate rivers and render them navigable. The sixteenth century was to see the start of a great reconquest of drowned and swamped land.

Notes

1. King Yt''Mr, the It'i-amara mentioned in an inscription of Sargon II (715 B.C.)
2. "Kūpa-yantra", see *Het Leemen Wagentje* (edit. Vogel, Amsterdam, 1897, 162)
3. "jala-yantra-cakra", see Böthlingk, *Indische Sprüche* No. 963.
4. Pāli "arghaṭṭa"
5. The sign PA_6, PA_5 (Šl 12), originally "crossroad" comes to mean "canal" (palgu, miṭertu), "ditch" (atappu), "conduit" (râṭu), small ditch (pattu).
   Another sign (F 260, B 61) picturing a field with ditches when combined with the sign for water becomes "inundated land".
   The expression "dyke and ditch" (iḳu u palgu) is a very common one.
   The sign temen (temennu) seems to depict a wooden fence held by three pickets, which served to strengthen a dyke. Šl 376, B 195, F 395). It came to mean "terrace, raised embankment". When combined with the sign for water we get (KAR, Kâru) dyke. The sign for inundate (RA, raḫâṣu) (B 273) shows a field with encirculing ditches. Such embankments were often the frontier between cities, like in the case of Lagash and Umma. The Sumerians distinguished between GAN-ID, the land along the river and GAN-GA, the land beyond. (Šl = Deimel, Sumerisches Lexikon; B = Burrows, Archaic Texts from Ur; F = Falkenstein, Archaische Texte aus Uruk)
6. The canal or dike repairers and builders (lúŠITIM-ÍD-DA, sikiru) as contrasted with the masons or architects (lúŠITIM), are mentioned apart from the "canal-opener" (mêḳiru) and the "canal-closer" (êkiru), these latter two seem to have been officials regulating the water supply. See S. Langdon, *Stud. Orient.* vol. I, 1927, 100—101; B. Meissner, *Lexicographisches IV*, M.V.Ae.G. vol. 15, 1910, 517.
7. "3ḫt" = Inundation
   "prt" = Coming forth (of the dryland)
   "šmw" = Lack of water (dry seasons)
8. Ordinary farming land = Eg. ḫrw, Gr. sporine gé
   Inundated land (by the river) = Eg. nḫb, Gr. bebregméne gé, Arab. rai
   Artificially irrigated land = Eg. kj.t, Gr. abrochos gé, Arab. sharaki
9. 'd mr = he who hacks the canal (open)
10. Chief or irrigation (ḫrp nd3.t.). Older titles of irrigation officials are:
    Chief of the "water-office" = ḫrp ḥt mw
    Chief of the canal-workmen = hrp šmsw mrw
    Scribe of the water-reservoir = šs š
    Inspectors of the inundation = šmsw mrj.t
    Minister of public works = mḏh nsw, lit. stonecutter of the King.

11. Province (city state) (Greek: nome) = sp.t, which like ḥsp (garden) is represented by a square complex of fields watered by crossditches.
"Dyke" (dnj.t, V.465.1) is derived from a verb "to hold back" (dnj, V.464.13). The dyke section that was cut was called ḥnw.t (III.110.7), the riverbank is called mrj.t (II.109. 12)
The "water-province" is also called ḳbḥw (V.29.5) which term later refers to the district of the first cataract.
"Canal" (šmw.IV.466.10) is derived from the verb "to go, to flow" (šm.t IV.466.1)
"Inundation" or "flood" (3gbj,I.22.10) is derived from the verb "to plant" (3g, I.22.8) hence 'śgb (IV.27.2) to water. The word for "river" or "Nile" (the only river!) (itr, I.146.10) actually means "inundation". Another word for inundation (ḥ3j, III.13.9, G.anabasis) gave rise to the word "mud" (ḥ3j.t, III.13.10) and "the irrigated part of the oases' (ḥ3j.t, III.13.11)
The "gate of a water ('3,I,164.24) may mean the mouth of a canal or a weir.

Bibliography

1. Ninck, M., *Die Bedeuting des Wassers im Kult und Leben der Alten* (Leipzig, 1921); Canaan, T., *Water and the "water of life"* (*J. Pal. O. S.* vol. 9, 1929, 57—69); Burrows, E. E., *The Use and Worship of Water among the Romans* (*Art and Archaeology* vol. 30, 1931, 221—228); Glover, R. T., *Springs of Hellas* (Cambridge, 1945); Grenier, A., *Manuel d'archêologie gallo-romaine*. IV. *Les monuments des eaux* (Paris, 1960);
Reymond, Ph., *L'eau, sa vie et sa signification dans l'Ancient Testament* (Vetus Test. Suppl. VI, Leiden, 1958); see also our *Studies Vol. I* (1964)
2. Corazza, O., *Geschichte der artesischen Brunnen* (Wien, 1901); Richter, J., *Zur vorgeschichtlichen Brunnenkunde* (*Mitt. Anthr. Ges. Wien*, vol. 53, 1923, 49); Ellis, A. J., *The divining rod* (*U.S. Geol. Survey*, Water Supply Paper 416, New York 1934); Bräunlich, E., *The well in ancient Arabia* (Leipzig, 1926); Tudor, D. & Bujor, E., *Die geheime Brunnenanlage von Sucidava* (Dacia IV, 1960, 541—552); Devoto, G., *Ausa le fonte* (Stud. Etr. XX, 1948/49, 151/7).
3. Sir Brown, A. H., *Irrigation, its principles and practice* (3rd Edit. London, 1920).
4. Etchwerry B. A., and Harding, S. T., *Irrigation practice and engineering* (2 vols. New York, 1933)
5. Audebeau Bey, C., *Les irrigations dans le monde antique* (*Revue Gén. des Sciences*, vol. 43, 1932, 272—282); Brittain, Robert, *Rivers, Man and myths* (New York, 1958)
6. Gompertz, M., *Corn from Egypt* (London, 1927)
7. Sir Willcocks, W., *Egyptian Irrigation* (London, 1899); id. *Irrigation in Mesopotamia* (London, 1917); Sir Brown, R. H., *Irrigation* (London, 1920); Stricker, B. H., *De overstroming van de Nijl* (Leiden, 1956)

8. Gordon Childe, V., *What happened in history* (London, 1942, 63—64)
9. Daryll Forde, D., *Habitat, Economy and Society* (London, 1934)
10. Rivers, W. H. R., *Irrigation and the cultivation of the taro* (In: *Psychology and Ethnology*, London, 1926)
11. Kosok, Paul., *The role of irrigation in ancient Peru* (*Proc. Eighth American Scient. Congress* vol. II, 169—178); Bushnell, G. H. S., *Peru* (London, 1960, 14, 15, 28, 44, 57, 58, 70, 89, 111, 114, 115)
12. Bryan, K., *Flood water farming in southern New Mexico* (*Geogr. Review* vol. 19, 1929, 453); Coe, M. D., *Mexico* (London, 1962, 103, 105); Braidwood & Willey *edit.*, *Courses toward Urban Life* (Edinburgh, 1962, 121, 122, 168, 169, 171, 183, 350); Millon, R., *Irrigation systems in the Valley of Teotihuacan* (Amer. Ant. 23, 1957, 2, 160—166); Nölle, W., *Indianische Bewässerunganlagen im Südwesten der Vereinigten Staten* (Gas. Wasserfach Dtsch. 98, 1957, 16, 384—387)
13. Schroeder, A. H., *Prehistoric canals in the Salt River valley, Arizona* (*Amer. Anthrop.* vol. 8, April 1943); Shetrone, H. C., *A unique prehistoric irrigation project* (*Ann. Rep. Smithson. Instit.* 1945, 379—386)
14. Dalman, G., *Arbeit und Sitte in Palästina*, Vol. II: *Der Ackerbau* (Gütersloh, 1932, 29); Vogelstein, H., *Die Landwirtschaft in Palästina zur Zeit der Mishnah* (Berlin, 1894, 10); Kedar, Y., *Water and soil from the desert. Some agricultural achievements in the ancient Negev* (Geogr. J. CXXIII, 1957, 179—187); Mayerson, Ph., *Arid Zone farming practices in Palestina tertia* (AJA 60, 1956, 180—181) (BASOR 153, 1959, 19—31)
15. Newman, J., *The agricultural life of the Jews in Babylonia between the years 200 C.E. and C.E. 500* (London, 1932, pags. 17, 40, 51, 82); Galling, K., *Biblisches Reallexikon* (1934); Hauck, A., *Realenz. f. Prot. Theologie* (3. Aufl.); Kraus, S., *Talmudische Archäologie* (vol. I, 1910, 78; Vol. II, 1911, 163)
16. Gen. 13.10; Deut. 8.7; Isa. 58.11; Prov. 5.15; Ezek. 31.14; Prov. 18.4
17. Prov. 17.14
18. Lamon, R. S., *The Megiddo Watersystem* (Chicago, 1935)
18a. Forbes, R. J., *Studies in Early Petroleum History* (Leyden, 1958, 174—179)
Norton, W. H., *Artesian Wells of Iowa* (*Iowa Geol. Surv.* VI, 122—124)
Bowman, I., *Well-Drilling methods* (*U.S. Geol. Survey Water-supply paper* no. 257, 1911)
19. ii Sam. 2.13
20. ii Sam. 4.12
21. Neh. 3.16; Eccl. 2.6
22. Dussaud, R., *La digue de lac de Homs* (*Monum. et Mém.* vol. 25, 1921/22, 133); Mazloum, S., *L'ancienne canalisation d'eau d'Alep* (*Docum. d'Etud. Orient. Inst. Franc. Damas* vol. V, 1936); Thoumin, R., *Note sur l'aménagement et la distribution des eaux à Damas et sa Ghouta* (*Bull. Etud. Orient.* vol. 4, 1934, 1—26); Tresse, R., *L'irrigation dans la Ghouta de Damas* (Paris, 1929)
23. Mercilek, E., *Beitrage zur Geschichte der Wasserwirtschaft* (*Wasserwirtschaft* vol. 24, 1931, 33—35; 89—90; 120—122)
24. Josephus, *Wars of the Jews*, I 9.4; *Antiquities* 18.3.2

25. MEANS, P. A., *Ancient civilisations of the Andes* (London, 1931); BUSHNELL, G.H.S., *Peru* (London, 1960)
26. TKATSCH, J., article "Saba" in the *Realenz. des Islams*; A. GROHMANN,, *Süderarabien als Wirtschaftsgebiet* (Prag. 1933, vol. II.19); RATHJENS-WISSMANN, *Süderarabienreise* (Hamburg 1934, vol. III.61); GLASER, E., *Reise nach Marib* (1913); RATHJENS, C., and WISSMANN, H. VON, *Vorislamische Altertümer*, pp. 145—158; MAZLOUM, S., *L'organisation hydraulique de deux oasis antiques: Qdevem et 'Amšreddi.*
27. SCHMIDL, M., *Die Grundlagen der Nilotenkultur* (*Mitt. Anthr. Ges. Wien*, 1935, 86—127)
28. FROBENIUS, *Das Unbekannte Afrika* (München, 1923, 60—62); LASSØE, J., *The irrigation system at Ulḫu* (*J. Cueiform Studies* vol. 5, 1951, 21—32); Millard BUTLER, A., *Irrigation in Persia by Kanáts* (*Civ. Engineer*, vol. 3, 69—73, 1933) TOLMAN, C. F., *Ground Water* (New York, 1937, 12—14); GOBLOT, H., *Le rôle de l'Iran dans les techniques de l'eau* (*Techniques, Art, Sciences* 1962, no. 155/6, pp. 7—18)
28a. LAESSØE, J., *People of Ancient Assyria, their inscriptions and correspondence* (London, 1963, 103—104)
On qanats in the Sahara see *Nat. Geogr. Mag CXIII*, 1958, 69ff and and *Science et Vie, Le Sahara* (1958) 58—61.
29. LUCKENBILL, D. D., *Ancient Records of Assyria and Babylonia* (Chicago, 1927, vol. II, 87—88); HERODOTUS, III.19; STRABO, XVI, 1.50, cap. 707
30. HAVELL, E. B. *History of the Aryan Rule in India* (London, 1918)
31. MACKAY, E., *The Indus Civilisation* (London, 1935, 174); MARSHALL, J., *Mohenjo Daro and the Indus Civilisation* (London, 1931, chap. X); SIR MORTIMER WHEELER, *India and Pakistan* (London, 1959)
32. MACKAY, E., *Early Indus Civilisations* (London, 1948, 120)
33. MEGASTHENES, XXXIV, 1
34. ARTHAÇASTRA, chap. 91
35. *Mahā- Bhāratā*, Sabhā. p. V 77
36. The "rājūkas", see BUHLER, *Z.D.M.G.* vol. 47, 466
37. The "nettika", see *Dhammapada*, proverb 80
38. BANNERJEE, N. C., *Economic Life and Progress in India* (Calcutta, 1925, Vol. I, 121); BARNETT, L. D., *Antiquities of India* (London 1913 102 & 107); BROHIER R. L., *Ancient irrigation works in Ceylon* (Colombo, 2 vols. 1934/36); TENNENT. J. E., *Ceylon* (London, 1860)
39. Société des Nations, *Rapport des Experts sur les Questions Hydrauliques et Routieres en Chine* (Genève 1936); CHÊNG TÊ-K'UN, *Prehistoric China* I (Cambridge, 1959, 111—113); BRAIDWOOD & WILLEY, *Courses toward urban life* (Edinburgh, 1962, 183)
40. MENG K'O, L. 126
41. CREEL, H. G., *The Birth of China* (London, 1936); EBERHARD, W., *Zur Landwirtschaft der Han Zeit* (*Ostasiatische Studien*, 1932, 74—105); HIRTH, F., *The Ancient History of China* (New York, 1923); GRANET, M., *Chinese Civilisation* (London, 1930)
42. LI, H., *Die Geschichte des Wasserbaues in China* (*Technikgeschichte*, vol. 21, 1931, 59—73); LI, H., *Über das Gon-Shue System so wie die Reglung des*

Huang-ho (Shanghai, *Wissenschaft* Bd. x, Heft 4); LOWDERMILK W. C. & WICKERS, D. R., *Ancient irrigation in China* (*Scient. Monthly* vol. 55, 1942, 209—225); HOMMEL, RUDOLF P., *China at Work* (New York, 1937, 49—54)

43. SHU CHUNG SHU, *Irrigation in ancient China* (In Chinese) (*Bull. Nat. Research Instit. Phil. Acad. Sinica*, Vol. V, 2, 1936)
44. MEHMKE, R. L., *Arbeitsgesinnung im Wandel der Zeiten* (Halbe 1930)
45. CREEL, H. G., *The Birth of China* (London, 1936); FORKE, A., *Der Ursprung der Chinesen* (Hamburg 1925)

45a. STEIN, A., *Ruins of Desert Cathay I*, 269, 376, 523ff
LE COQ, A. VON, *Auf Hellas Spuren in Ost-Türkestan* p. 105.

46. HERODOTUS, I.193
47. *Enquiry into Plants*, VIII. 7.4
48. STRABO, *Geography* XVI. 1.14. cap. 742
49. Jeremiah 51.13, 51.42—43, 51.36
50. SPRENGER, A., *Babylonien, das reichste Land der Vorzeit* (Berlin, 1886)
51. SIR WILLCOCKS, W., *Irrigation in Mesopotamia* (2nd Edit. London 1917); Herm. WAGNER, *Die Überschätzung der Anbaufläche Babyloniens und ihr Ursprung* (*Nachr. K. Sächs, Ges. Wiss. Gött. Phil.-hist. Kl.* 1902)
52. STRABO, XVI. 1.9.10, cap. 740
53. HERZFELD, E., *Über die historische Geographie von Mesopotamien* (*Petermanns Mitt.* vol. 55, 1909, 345)
54. MACKAY, D., *Ancient River Beds and Dead Cities* (*Antiquity* vol. 19, 1945, 135—144); DELATTRE, A., *Les travaux hydrauliques en Babylonie* (*Rev. Quest. Scient.*, 1888, 476—507); MEISSNER, B., *Babylonien und Assyrien* (Heidelberg, 1923, vol. II, 375)
55. BEWSHER, J. B., *On part of Mesopotamia* (*J. R. Geogr. Soc.* vol. 37, 1867, 160—182); HERZFELD, E., *Mythos und Geschichte* (*Iran. Forschungen* vol. VI, 1934, 42); HERTZ, A., *Die Kultur um den persischen Golf* (*Klio Beihefte* vol. XX, 1930)
56. AUDEBEAU BEY, C., *Les irrigations dans le monde antique* (*Rev. Gén. des Sciences* vol. 43, 1932, 272—282); GRUYTER, P. DE, *Drainage in Iraq* (*De Ingenieur* 1953 B.1 B.17)
57. DOSSIN, G., *Le dieu Gibil et les incendies de Végétation* (*Rev. Hist. des Relig.* 1934, 28—62)
58. FISH, T., *Aspects of Sumerian Civilisation during the Third Dynasty of Ur. III: Rivers and canals* (*Bull. John Rylands Libr.* vol. 19, 1935, 98); SCHNEIDER, A., *Die sumerische Tempelstadt* (Berlin, 1920, 43); GRUBER, J. W., *Irrigation and Land Use in Ancient Mesopotamia* (*Agricultural History* vol. 21/22, 1947/48, 69—77); IONIDES, M. G., *Two ancient irrigation canals in Iraq* (*Geogr. J.* vol. 92, 1938, 351—354); KUPPER, J. R., *Le canal Išim-laḫdunlim* (*Bibl. Orient.* vol. IX, 1952, 168—169)
59. BARTON, *Inscriptions of Sumer and Akkad* (New Haven, 1929, 14, 57—61)
BRINKMAN, J. A., *Provincial administration in Babylonia under the second dynasty of Isin* (JESHO VI, 1963, 233—242)
60. WOOLLEY, L., *Development of Sumerian Art* (London, 1935, 112 & Plate 65a)
61. FISH, T., *Aspects of Sumerian Civilisation during the Third Dynasty of*

Ur. III: *Rivers and Canals* (*Bull. J. Rylands Libr*. vol. 19, 1935, 98)
61a. EDZARD, D. O., *Die "zweite Zwischenzeit" Babyloniens* (Wiesbaden, 1957, 112—117)
LAESSØE, J., in *JCS* V, 1951, 25
KRAUS, F. R., *ZA* 55, 1955, 52ff.
KUPPER, J. R., *RA* 53, 1959 28ff., 36ft.
WASCHOW, *MAOG* X, 1, 44—61
ARCH. R. DE, MARI, vol III, 1—11, 75—80
KUPPER, J. R., *RA* 41, 1947, 175ff.
MDP XXII, 58ff & 89ff; *MDP* XXVIII 398, 452, 453
62. JACOBSEN T. & SETON LLOYD, *Sennacherib's aqueduct at Jerwan* (Chicago, 1935) (See also my Studies, Vol. I)
63. FUAD SAFAR, *Sennacherib's Project for Supplying Erbil with Water* (Sumer 1947, 23—25)
64. STRABO, *Geography* XVI, 1.11, cap. 7.11
65. ARRIAN, *Anabisis of Alexander* VII, 21, 1—5
66. POLYBIUS, IX.43
66a. GRUYTER, P. DE, *De Ingenieur* 65, 1953, B. 1—16; ADAMS, R. M., *Akten des* 24. *Int. Orient. Kongresses* 1957, 139ff; JAKOBSEN, TH. and ADAMS, R. M., *Salt and silt in ancient Mesopotamia* (Science (USA), 1958, 1251—1256)
67. STRABO, XVII, 1.3 cap. 788
68. Chapters 6 and 125 of the so-called *Book of the Dead*
69. AUDEBEAU BEY, C., *Les irrigations dans le monde antique* (*Revue Gén. Sciences* vol. 43, 1932, 272—282); OBERMEYER, J., *Die Landschaft Babyloniens in Zeitalter des Talmuds und des Goanats* (Berlin, 1929, 52 & 118)
70. BREASTED, J. H., *Ancient Records of Egypt* (3rd edit., Chicago, 1927 vol. IV, 726) (to be cited as AR)
71. BORCHARDT, L., *Nilmesser und Nilstandsmarken* (Berlin, 1906/1934, 2 vols.); id., *Zu den bei Gebel Silsile neugefundenen Nilmessern* (*Z. Ae.S*. vol. 72, 1936, 137—139); id., *Nachträge zu Nilmesser und Nilstandsmarken* (*Sitzber. Preuss. Akad. Wiss. phil. hist. Klasse*, 1934, 11—13, 194—202); POPPER, W., *The Cairo Nilometer* (Berkeley (Calif.), 1951); SETHE, K., *Untersuchungen* III (Berlin, 1905, 103; DRIOTON, E., *Les origines pharaoniques du Nilomètre de Rodah* (*Bull. Instit. Egypte*, vol. XXXIV, 1950, pags. 291—316); ENGREEN, F. E., *The Nilometer in the Serapeum at Alexandria* (*Medieavalia et Human*. vol. I, 1943, pags. 3—13); GALEB PACHA, K. O., *Le Mikyâs ou Nilomètre de l'île de Rodah* (*M.I.E*. vol. 54, 1951)
72. *A.R*. vol. I, 90—167
73. DIODORUS, I.36
74. PLINY, *Nat. History* V.10
75. BRUNNER, H., *Eine Kalksteinscherbe... einer Nilüberschwemmung* (*Z. Aeg. S*. vol. 76. 1940, 1—3); JANSSEN, J., *Bemerkungen zur Hungersnot im alten Aegyten* (*Biblica* vol. 20, 1939, 69—72)
76. STRABO, *Geography* XVII, 1.48, cap. 717
77. HERODOTUS, II.97
78. *AR* IV. 266

79. *AR* III.275; see also Smither, P. C., *A Tax-assessor's Journal* (*J.E.A.* vol. 27, 1941, 74—76)
80. Herodotus, II.109; Strabo, XVII.1.3, cap. 787
81. *AR* I.93, 118
82. Suz. Borger, *A note on some scenes of land measurement* (*J.E.A.* vol. 20, 1934, 54—56); Borchardt, L., *Statuen von Feldmessern* (*Z. Ae.S.* vol. 42, 1905, 70—72)
83. Strabo, XVII.1.37, cap. 811 "kleithra"; see also Pearl (*Aegyptus* vol. 31, 1951, pags. 223—230)
84. Daressy, G., *L'eau dans l'ancienne Egypte* (Cairo 1915); Spiegelberg, W., *Zwei Ausdrücke der Bewässerung* (*Z. Ae.S.* vol. 53, 1917, 113); Newberry, *Beni Hasan* Vol. I, plate 11929; el. Bersheh plate 26; Garris Davis, N. de, *The Tomb of Nefer Hotep at Thebes* (New York 1934)
85. *A.R.* I.407
86. *A.R.* I.324
87. *A.R.* I.642—648
88. *A.R.* II.75; III.649
89. Posener, G., *Le canal du Nil à la Mer Rouge* (*Chronique d'Egypte* vol. 13, 1938, 258—298); Bourdon Cl., *Anciens canaux, anciens sites et ports de Suez* (Caire, 1925)
Herodotus, II. 158; Strabo, XVII.1.25. cap. 804; Pliny, *Nat. Hist.* VI.145
90. Herodotus, II.4 & 99
91. Brown, R. H., *The Fayum and Lake Moeris* (London, 1892) Sir Flinders Petrie, W., *Social Life in Ancient Egypt* (London 1923)
92. Caton Thompson G. & Gardner, E. W., *Recent work on the problem of Lake Moeris* (*Geogr. J.* vol. 73, 1928, 20); Caton Thompson, G., *The Desert Fayum* (London, 1934, 2 vols.); Caton Thompson, G., *Kharga Oasis* (*Antiquity* vol. 5, 1931, 221—226); Gardner, E. W., *The origin of the Fayum Depression* (*Geogr. J.* vol. 74, 1929, 371—383); Sandford K. S. & Arkell, W. J., *The origin of the Fayum Depression* (*Man*, April 1929, 578—584); Pochan, A., *Note sur le sujet de la gorge d'Illahum et le lac Moeris* (*B.I.E.* vol. 18, 1935/36, 131—136)
93. Murray, G. W., *Dessication in Egypt* (*Bull. Soc. Roy. Géogr.* 1949, 19—34)
94. *A Dispute over Suicide*, line 65 (In: *Ancient Near Eastern Texts*, Princeton, 1950, 405)
95. Eyth, Max, *Das Wasser im alten neuen Aegypten* (In: "*Lebendige Kräfte*, Berlin, 1905, 27—75); Mehmke, R. L., *Beiträge zur Geschichte des Wasserbaus im alten Aegypten* (*Technikgeschichte* vol. 16, 1926); Mehmke, R. L., *Arbeitsgesinnung im Wandel der Zeiten* (Halle, 1930)
96. Boak, A. E. R., *Irrigation and Population in the Fayum* (*Geogr. Rev.* 1926, 353); Boak, A. E. R., *Notes on canal- and dike-work in Roman Egypt* (*Aegyptus* vol. 7, 1926, 215—219); Calderini, A., *Recerche sul regime delle acque nell'Egitto greco-romano* (*Aegyptus* vol. II, 1920, 37 & 189); Hartmann, F., *L'agriculture dans l'ancienne Egypte* (Paris 1923); Johnson, A. C., *Roman Egypt* (Baltimore, 1936); Lozach, J., *Le delta du Nil* (Paris 1936); Pearl, O. M., *Irrigation works on 3 canals in*

the Arsinoite nome (*Aegyptus*, vol. 31, 1951, pags. 223—230); WADDELL, G., *On Egypt by Aristides of Smyrna* (*Bull. Arts*, Univ. of Egypt, Europ. Sect. vol. II, 1934, 121—166); WESTERMANN, W. L., *The dry land in Ptolemaic and Roman Egypt* (*Class. Phil.* 1922, 21); id., "*Inundated Land" in Ptolemaic and Roman Egypt* (*Class. Phil.* 1920, 120; 1921, 169); id., *The irrigation system of Egypt* (*Class. Phill.* 1919, 158); id., *Aelius Gallus and the reorganisation of the irrigation system under Augustus* (*Class. Phil.* 1917, 237); id., *Dike corvée in Roman Egypt* (*Aegyptus* vol. 6, 1925, 121—129); YEIVIN, S., *The Ptolemaic System of Water Supply in the Fayum* (*Ann. Serv. Antiq. Egypte* vol. 20, 1930, 27—30); SCHNEBEL, M., *Die Landwirtschaft im hellenistischen Aegypten* (München, 1925); VARCL, L., *Zum Bewässerungswesen im römischen Aegypten* (*Archiv. Pap. forsch.* XVII, 1960, 17—22)

97. BARROIS, J., *Irrigation in Egypt* (Washington, 1889)
98. BOUCHÉ-LECLERQ, *L'ingénieur Cléon* (*Rev. Etud. Grecq.* 1908, 121)
99. *Pap. Lille* I
100. *Pap. Petrie* III.37 & 40
101. *C.I.G.* 4863; BALTY-FONTAINE, J., *Pour une nouvelle édition du Liber Aristotelis de Inundacione Nili* (*CdE* XXXIV, 1959, 95—102); PEARL, O. M., *The inundation of the Nile in the second century A.D.* (*Trans. Amer. Phil. Assoc.* LXXXVII, 1956, 51—59
102. *S.B.* 7361 of A.D. 211/212
103. VITRUVIUS, X.5
104. VITRUVIUS, X.7
105. LITTMANN, E., *Die sâqiya* (*Z.Ae.S.* 76, 1940, 45—54)
106. WIEDERMANN, E., *Beiträge z. Gesch. der Naturwissenschaften, X* (*Sitz. ber. Phys. med. soz. Erlangen* 38, 1906, 307—340); COLIN, G. S., *La noria marocaine* (Hespéris 14, 1932, 22—49); LAUFER, B., *The noria or Persian wheel* (*Oriental Studies in honour of C. E. Pavry*, Oxford Univ. Press, 1933, pags. 238—250); SCHIOLER, TH., *Virkningsgraden af en maurisk noria på Ibiza* (*Ingeniør og bygings vaesen* 56, 1961, 261—268); SCHIØLER, TH., *Øsevaerket* (*Naturens Verden* (Copenhagen, July 1963, 209—219)
107. CHATLEY, H., *Engynes* (*Engineering* 1946, p. 388 etc.); YOUATT, C. S., *Methods employed in raising water* (*Trans. Manchester Ass. Eng.* 1945/46, 3—20)
108. *Mém. Dél. en Perse* vol. VII, p. 180 & fig. 472; AUGAPFEL, *Bab. Rechtsurkunden*, p. 63
109. SPIEGELBERG, W., *Zwei Ausdrucke der Bewässerung* (*Z.Ae.S.* 53, 1917, 113); For a picture of an early Egyptian shadoof see QUIBELL & GREEN, *Hierakonpolis*, vol. II, 1902, plates 74 & 75.
110. PATERSON, *Palace of Sennacherib*, pl. 32 f.
111. WARD, *Seal Cylinders*, 1910, fig. 397; DELAPORTE, *Cat. Cylindres Orient. du Londre*, *Vol. Acquis*, No. A—156
112. VITRUV, X.1
113. STRABO, XVI.1.5. cap. 738; see also EBELING, E., *Akkadische bakrâtu = Rad einer Bewasserungsmachine* (*Orientalia*, vol. 20, 1951, pags. 13—14; LAESSØE, JORGEN, *The meaning of the word alamittu* (*CRRAI* III, 1954,

150—156); LAESSØE, J., in *JCS* VII, 1953, 5—16; LAESSØE, J., *The Shemshâra Tablets* (Copenhagen, 1959, 69)
114. DIODOR, II.1
115. *P. Cor.* 5
116. *B.G.U.* 1120,27
117. STRABO, XVII.1.30. cap. 807
118. CALDERINI, A., *Macchine idrofore secondo i papiri greci* (*Rend. R. Ist*, Lombardo ser. II, vol. LIII, 1920, 620—631)
119. *Pap. Oxyr.* 971
120. DIODOR, I.34
121. DIODOR, V.37
122. VITRUV, X.6
123. PRICE, *Proc. Soc. Antiquaries*, 2e series vol. XVI, p. 277
123a. FORBES, R. J., *Studies in Ancient Technology* Vol. VII, 1963, 212
124. *Pap. Lond.* 1177
125. *P. Oxyr.* 707
126. *P. Oxyr.* 729
127. *P.S.J.* 921
128. DALMAN, G., *Arbeit und Sittein Palästina* (Vol. II, Gütersloh 1932, p. 222)
129. EGGERS, G., *Wasserversorgungstechnik im Altertum* (*Technikgeschichte* Bd. XXV, 1936, p. 1—13); WIEDEMANN, E. & HAUSER, F., *Über Vorrichtungen zum Heben von Wasser in der islamischen Welt* (*Beitr. Gesch. Ind.* Bd. VIII, 1918, p. 121); SCHMELLER, H., *Beiträge zur Geschichte der Technik in der Antike und bei den Arabern* (*Abh. z. Gesch. d. Naturwiss. u. d. Med.* Bd. 6, 1922)
130. ASHBY, TH., *The Aqueducts of Rome* (Oxford, 1935, 46—47)
131. PLINY, *Nat. Hist.* XVII.250; HARMAND, L., *Les terres et ses problèmes dans l'antiquité grecque et romaine* (*Inform. Hist.*, Paris, XVIII, 1956, 59—68); WHITE, K. D., *The efficiency of Roman farming under the Empire* (*Agric. Hist.* 30, 1956, 85—89)
132. BRESCH, R., *L'eau dans le monde romain* (*L'Eau*, vol. 25 1932, 3 C); BRUNHES, J., *L'irrigation... dans le péninsule ibérique et dans l'Afrique du Nord* (Paris, 1902); JAUBERT DE PASSA, *Recherches sur les arrosages chez les peuples anciens* (Paris, 1846/47, 4 vols); Ch. KNAPP, *Irrigation amongst the Greeks and Romans* (*Class. Weekly* vol. XIII, 73—74, 81—82); GLOVER, R. T., *The Greek Farmer & The Greek and the Forest* (In: *The Challenge of the Greek*, Cambridge, 1943)
133. *C.I.L.* VIII. 18587; VITA-FINZI, CL., *Romans Dams in Tripolitania* (*Antiquity* XXXV, 1961, 14—20)
134. ANDRÉ KENNY, E. J., *The ancient drainage of the Copais* (*Ann. Archaeol. Anthr.* 1935, 189—206); KAHRSTEDT, U., *Der Kopaissee im Altertum und die Minyschen Kanale* (*J. D. Arch. Instit.* vol. 52, 1937, 1—20)
135. THEOPHRASTUS, *Hist. Plant.* IV, xi, 2; *De Caus. Plant.* xii.3
136. HOMER, *Iliad* V.707; STRABO, IX. cap. 415
137. PAUSANIAS, IX.38.7
138. *I.G.* VII.2712; VII.2792
139. STRABO, IX.2.18 cap. 407; DIOG. LAERTIUS, IV.4.6; STEPHANUS of

Byzantium, *Athenai* No. 8
140. Diogenes Laertius, VIII, 51—77
141. *I.G.* XII.9.191
142. Leger, A., *Les Travaux Publics aux temps des Romains* (Paris, 1875); Merckel, K., *Die Ingenieurstechnik im Altertum* (Berlin, 1899); Korthalt Altes, J., *Polderland in Italie* (The Hague, 1928, 1—12); Gest, A. P., *Engineering* (Londen, 1930); Gall, J. le, *Le Tibre dans l'Antiquité* (Paris, 1943); Rodenwaldt E. & Lehmann, H., *Die antiken Emissare von Cosa-Ansedonia, ein Beitrag zur Frage der Entwässerrung der Maremmen in etruskischer Zeit* (*Sitz. ber. Heidelberg Akad. Wiss.* 1962, 1.
143. Suetonius, *Claudius* 20, 1—2; 21,6; Tacitus, *Ann.* XII. 56—57; Pliny, *Nat. Hist.* 36.124; *C.I.L.* IX.3915; Cozzo, G., *Ingegneria Romana* (Rome 1928, 310—317)
144. Livy, *Epit.* 46
145. Suetonius, *Julius* 44.3
146. Suetonius, *Nero* 32.3
147. Strabo, V.3.6. cap. 233
148. Horace, *Satires* I.V.12
149. Strabo, V.1.11. cap. 217
150. Averone, A., *Sull'antica idrografia veneta* (Mantua, 1911, 156); Barnabi, F., *Notiz degli Scavi* 1915, 137—144
151. Ghirardini, *Notiz. degli Scavi* 1921, 24—36
152. Buongiorno, A., *Le Bonifiche italiana*, 27
153. Strabo, V.1.7 cap. 213; Jordanes, *Getica* 29.150
154. Plutarch, *C. Marius* 15
155. Panegyr. V (VIII).6.2—3
156. Tacitus, *Annals* I.6; II.8; XIII.53; *Hist.* V.19; Suetonius, *Claudius* I
157. Hettema, H., *De Nederlandsche wateren... in den Romeinschen tijd* (The Hague, 1938, 105); Vollgraff, C. W., *De dijk van Drusus* (*Ned. Kon. Akad. Wet. A'dam*, vol. 1, 1938, 12, 555—576); Vollgraff, C. W., *De moles van Drusus* (*Ned. Kon. Akad. Wet. A'dam* vol. 2, 1939, 6, 141—143)
158. Tacitus, *Annals* XI, 20.2
159. Veen, Joh. van, *Dredge, Drain, Reclaim* (The Hague, 1955); Bijl, J. G., *Au pays des polders* (*Congres Int. de Géographie, Amsterdam*, 1938); Cools, R. H. A., *Strijd om den grond in het lage Nederland* (The Hague, 1948); Dibbits, H. A. M. C., *Nederland - Waterland* (Utrecht, 1950); Korthals Altes, J. *De eerste bedijking der groote en kleine Moeren in West-Vlaanderen* (Brugge, 1924); Hérubal, M., *L'homme et la côte* (Paris, 1936)
160. Pliny, *Nat. Hist.* XVI.1
161. Dante, *Inferno*, Canto 15
162. Tacitus, *Annals* 13.57
163. Thierry, J., *De strijd tegen Nederlands erfvijand* (*De Ingenieur*, vol. 45 1930, A 47—A 55); Visser, M. F., *Bedijkingen voorheen en thans* (The Hague, 1941); Möhlmann, *Die Entwicklung des niederlandischen Deichwesens im Laufe der Jahrhunderten* (*Z.f.Bauwesen* vol. 80, 1930, 259—264)

ANDREAE, S. J. FOCKEMA, *Embanking and drainage authorities in the; Netherlands during the Middle Ages* (*Speculum* 27, 1952, 158—167)
164. VIERLINGH, ANDRIES, *Tractaet van Dijckagie*, 1570 (The Hague, 1920)
165. SGANZIN, J. M., *Programme ou résumé des leçons d'un cours de construction...* (Paris, 1823)
166. HÉRUBAL, M., *L'homme et la côte* (Paris, 1937, 109); DOORMAN, G., *Octrooien van Uitvindingen in de Nederlanden* (16e—18e eeuw) (The Hague, 1952, Supl. vol. page 24—25); DOORMAN, G., *Techniek en Octrooien*, The Hague, 1953, 81
167. CONRAD, F. W., *Bijdrage tot de geschiedenis der eerste schutsluizen in Holland* (*De Vriend des Vaderlands* 1831, 748—764); *Verspreide Bijdragen*, The Hague 1849, 3)
168. STROOTMAN, J., *Over wijde zeesluizen en sluisdeuren uit plaatijzer* (*Verh. Kon. Instit. v. Ing.* 1863/64, 12)
169. KIRBY, R. S. & LAURSON, PH. G., *The early years of modern civil engineering* (New Haven, 1932, 52—53); UCCELLI, A., *Storia della Tecnica* (Milano, 1945, 338—345); PARSONS, W. B., *Engineers and in the Renaissance* (Baltimore, 1939, 372—398); BRITAIN, R., *River technology and historical development* (*Congrès Int. Hist. Sci.*, 1962)
170. *Codice Laurenziano* No. 361, Florence
171. *Codex Atlantico* 7 v.b., 240 r—c; 33 v—a; 151 v—b; 341 v—b
172. ZONCA, V., *Novo teatro di machine et edificii* (Padua, 1607)
173. STEVIN, SIMON, *Nieuwe Maniere van Sterctebou door Spilsluysen* (bound with his *Castrametatio*, Rotterdam, 1617) (The Principal Works..., Vol. V, Amsterdam. 1965
174. *Douai archives* Nos. DD 20, 30 & 439
175. LOOSJES, A., *Onze Windmolens en hunne geschiedenis* (*Vragen v. d. Dag* vol. 35, 1920, 401—416; 485—497); BOONENBURG, K., *Onze Windmolens* (Amsterdam, 1949)
176. CONRADIS, H., *Alte Baggermaschinen* (*Technikgeschichte* vol. 26, 1937, 51—61); DOORMAN, G., *Hollandsche oude Baggermolens* (*De Ingenieur* 38, 1951, 38, 1—5)

CHRONOLOGICAL SURVEY OF IRRIGATION AND DRAINAGE DATA

3500 B.C.	Irrigation spreads in the Ancient Near East, strong increase of population.
3000	Basin irrigation in Egypt, Perennial irrigation in Mesopotamia, "Department of irrigation" and land registry office in Egypt, Nilometer at Memphis.
2500	Ur-Nanše of Ur digs irrigation canals.
2400	E-Anna-tum digs the water reservoir Luman-dimdug.
2400	Entemena of Lagaš digs the Shatt-el-Hai and other canals.
2400	Merenre builds a shipping canal at the first cataract of the Nile.

2280	Rice-irrigation in the Hoang-Ho valley, China.
2200	Fayum works of Amenemhat II, Nilometer at Elephantine, Egypt. Shadoof in use in the Near East. Irrigation canals in China.
2100/1950	The Third Dynasty of Ur reshapes and rebuilds the canal-system.
1950	Sesostris I builds Nile-Red Sea canal.
1875	Seostris III rebuilds first cataract canal (260′ long, 35′ wide, 26′ deep).
1800	Seti I redigs the Red Sea canal.
1750	Active canal building by Sin-idinnam and Rim-Sin of Larsa.
1700	Chammurabi of Babylon builds canals and codifies irrigation laws.
1400	Drainage canals of Lake Copais built by the Minyans of Orchemenos.
1250	Ramses II rebuilds the Nile-Red Sea canal.
1125	The Zalmani canal (n.e. of Babylon) built.
1100	The Chou Li, first Chinese handbook on hydraulic engineering.
1000	Oldest weirs of India.
750	Lokman builds valley-dam of Marib (S. Arabia). Irrigation near Damascus.
700	Sargon II builds the "canal of the land of Umliaš" at the Persian border. Sennacherib's irrigation canals for Niniveh, Chorsabad and Arbela.
650	River King and River Lo (China) connected by canal by Prince of Ch'in.
620	First wooden bridge at Rome.
620/540	Neobabylonian kings reshape canal-system, drainage of "sea land" at Lower Euphrates.
600	Necho II of Egypt rebuilds the Nile-Red Sea canal. Periandros constructs "diolkos" (wooden strip of rollers) to convey ships over the isthmus of Corinth.
540/523	Harbour of Samos built.
522/520	Darius I, reconstructs Nile-Red Sea canal and opens trade between Egypt and India.
c. 500	Barrage of Abhaya on Ceylon. Cloaca Maxima built at Rome.
480	Xerxes builds canal of Akanthus and ship-bridge over Hellespont.
469/426	Harbour of Piraeus built.
460	Anaxagoras explains risings of the river Nile.
450	Empedocles drains swamps of Selinus, Herodotus mentions a 400 m mole at Samos and a diving-apparatus.
410	Well-theory of Hippocrates.
c. 400	Si-men-Pao drains and irrigates district of Ho-ni between Yellow River and River Chang, China.

396	Tapping of Lake Albano by 1800 m tunnel (area 2 × 1.10 m).
350	War-harbour of Athens. Aristotle describes the diving bell.
330	Aristotle's theory of the life-cycle of water (rains, sources, rivers, sea, evaporation).
c. 315	Chaerephanes drains marshes near Eretria (Euboea, Greece).
307/261	Harbour of Seleucia built.
c. 300	Crates builds canal in Copais plain. Roman sewer system extended. Weirs in Ceylon. General introduction of water-wheels and Archimedean screw in Egypt and gradually in Near East.
289	Deepening of Lake Velinus and drainage of plain of Raete.
285	First lighthouse, Pharos, of Alexandria.
280	Ptolemy II finishes Necho's Red Sea canal.
272	Attempt at drainage of Pomptine marshes south of Rome.
260	Cleon, chief irrigation officer of the Fayum reclaims 2700 ha of land.
225	Shih-Huang-Ti connects rivers Ch'i and Huai by canal.
218	Harbour of Tarraco built.
160	Pomptine marshes drained by consul Cegethus.
115	Cimbrian flood in north-western Europe. Canals built to connect the basins of the Si-Kiang and Yangtzekiang, the Wei and Hoangho, and the Han and Yangtzekiang (China).
109	M. Aemilius Scaurus builds drainage canals at Placentia between Parma and the river Po.
104/102	Gaius Marius digs the army canal, Fossa Marina from Rhone-Marseilles.
50	Caesar plans drainage of Lake Fucine and Pomptine marshes.
35	Harbour of Baia started by Agrippa.
31	Adige canalised.
12 B.C.	Drusus' canals and dykes in Holland (Rhine and Yssel connected).
	Canalisation of branch of Po to Ravenna for fleet station.
41/52 A.D.	Drainage of Lake Fucine by 5640 m tunnel.
42	Lighthouse and new harbour of Ostia at mouth Tiber.
c. 45	Seneca's theory of wells.
	23 mile canal between Rhine and Meuse (Holland) dug by Corbulo.
60	Nero attempts to cut the isthmus of Corinth and to build a canal between Ostia and Lake Avernus (Puteoli) but the work is stopped at Misenum and never continued to Ostia.
100	Lighthouse of Coruna (Spain).

101/104	Trajan builds the hexagonal inner harbour (Portus) of Ostia.
115	Cleaning and reopening of the Fucine tunnel.
250	Probus regulates the river Nile.
276/282	Drainage of the valley of Sirenus.
500	Climate of n.w. Europe becomes wetter and colder (Subatlantic phase).
500/900	Artificial mounds (terpen, wierden) built in Friesland and Groningen.
600/800	The first dikes constructed in the Low Countries.
600/610	The 1000 km Imperial Canal (Hoangho to Yangtzekiang) started.
645	Nile-Red Sea canal reopened.
719	Nilometer of Rhoda (Cairo) built.
756	Arabs start systematic irrigation of Spanish rivervalleys.
793	Charles the Great constructs a canal from Rezat to Altmühl.
802	The Lex Frisionum.
839	Storms and floods inundate part of the Low Countries.
950	Timber trade begins on the Elbe and Moldau.
1000	Plans for a weir in the Nile.
1065	Flood-gate at Krooswijk (Holland).
1066	Jade-bay in n.w. Germany formed by storms and floods.
1077	Arnoriver regulated within the city bounds of Florence.
1106	Dutch colonists reclaim marshes north-east of Bremen.
1115	The city of Bremen obtains the right of buoyage of the Weser.
1135/1146	Donau bridge built at Regensburg.
c. 1150	Rüstinger Laws of East Friesland defining cooperation in dike-building codified. Frisian XVII general ordnances mention dike-building. Formation of drainage units (Hoogheemraadschappen) in the Low Countries.
	Benedictines and Cistercians engaged in drainage of swamps and lakes. Use of refuse- and drain-water on meadows near Milan.
	Beginnings of canal- and gate-building. Weirs with inclined planes and lifts.
1164	St. Julian flood in the North Sea.
1177/1183	The 16 mi. Naviglio Grande from Ticino to Po.
1200	Sachsenspiegel on the "law of the dike".
1234	Wooden chamber-lock in Zwin near Damme (Flanders).
c. 1250	West-Frisian dike built in North Holland province. Clay dike with sand core introduced.
1277	Dollart and Zuiderzee formed by floods.
1285	Lift-lock at Spaarndam (Holland) built.
1286	Lighthouse built at Elbe mouth.
1287	Flood in Friesland and Ems district (Dec. 14).
1300/1400	Middelzee in Friesland reclaimed.

	Introduction of the sea-weed dike and the reed-dike.
	First reclamations on Oder and Wistula.
	Flood-gates in general use in Holland.
1319	West-Frisian dike completed.
1344	Edict of Count William IV of Holland mentions "windmills used for flour production and waterpumping".
1350	Canals and dikes built in Bohemia.
1359	Galeazzo Visconti begins the canal from Milan to Pavia.
1366	Lift in weir near Schwellingen in the Saal river.
1393/1398	Ship's canal of Stecknitz with lifts built.
1394	The Zwin lift-lock rebuilt.
1395	A gate built in the Naviglio Grande of Milan.
c. 1400	Naviglio Grande connected with city moat of Milan.
	Wind-driven scoop-wheels gradually in common use for drainage (Holland) (Tower mills)
	Pallisade protection of dikes introduced.
1404	Floods in the province of Zeeland drown part of the islands.
1408	Marshmills tested near Alkmaar.
1421	St. Elizabeth's Flood destroys part of province of Holland.
1430/1450	Drainagemills (wooden) of the "wipwatermolen" type built.
1433	Lift-lock at Vreeswijk (Holland) in use for ships.
1438	Primitive dredger ("krabbelaar") used in Middelburg, Zeeland.
1450	Single gates at intervals along a river in common use in Italy.
1450/1568	Centralisation of drainage policy in the Netherlands.
1452	Leon Batista Alberti invents chamber-locks.
	Bertola da Novate constructs Bereguardo canal (at least 12 locks); Milan-Pavia (5 locks).
1457	Bertola da Novate builds Martesana canal from Lake Como and Adda to Milan, two locks.
1471/1542	Lauwerszee in Friesland reclaimed.
1475	Drainage mills with Archimedean screws in use in Holland.
1494	Leonardo da Vinci becomes Ingegnere Camerale of Ludovico SFORZA.
1495	First dry-dock at Portsmouth.
	Leonardo plans control and regulations of Arno with weirs and lifts for ships. Invents mitered gates and small wickets.
1500/1600	First experiments with stone-clad dikes.

Chapter two

POWER

Introduction

At each stage of its history technology is the resultant of many interacting factors such as the available materials, the accumulated skill and experience of arts and crafts of the day, economic and social conditions, religious and ethical tenets and philosophical doctrines. At each stage the craftsmen dispose of a certain amount of tools and machines to execute the various operations needed to transform their raw materials into the products needed. Such things as the hammer and chisel or the rotary quern are sometimes called "direct actors" to distinguish them from the type of machinery that supplies the motive power for other tools or machinery. This latter type, called *prime movers*, represent the stage of harnessing natural forces which mankind has by then attained. They convert the energy of animal muscles, running water, wind or heat into motive power for running other machinery.

Therefore at each stage the availability of such prime movers is the keystone of technology. The prime mover determines the size of the units (lumps of metal, wood, etc.) the craftsmen can work and at the same time the size of the machinery, tools and products they can finally make from the units thus fashioned (1). The introduction of a new prime mover generally means a more concentrated form of energy, a new level at which things can be made and produced, which upto then were only potentially possible and never realised. The new prime mover means the realisation of the dreams of earlier generations of craftsmen and engineers.

Written from this point of view the history of technology would show five stages. During the first stage mankind disposes of the energy of its muscles only. With the domestication of animals during the New Stone Age mankind merely increased the total amount of tractive power available without raising the level of energy production for reasons to be discussed later on. Even the advent of the horse and the camel during the second millennium B.C. did not change this situation. The second stage is, therefore, important mainly for this increase in total man- and animal-power rather than for a higher level, a more concentrated form of energy production. Still its results are manifest in the greater variety and larger number of tools produced from the earliest days of the Ancient Empires onwards. Man— and animal—power do not lead to large machinery.

The third stage opens with the introduction of the water-mill during the Later Roman Empire. Its earliest form, the Norse mill, again means a mere shift of the motive power from animal muscles to a machine moved by running water rather than a new level of energy production. The querns movel by two slaves or the donkey-mills of 0.4—0.5 HP were now sometimes moved by a primitive mill, the energy output of which was hardly more than 0.5 HP. This primitive machine meant a new addition to the total energy available to mankind but not a new level of energy available.

When, however, the Roman engineers converted this primitive mill into the Vitruvian mill, which we will discuss in detail later on, they created a prime mover, which even in its most primitive form represented an output of some 3 HP. For several reasons (see p. 98) this water-mill did not come into general use in the Mediterranean world. Once Western Europe realised its great importance in the early Middle Ages its rapid introduction and technical development (see p. 106) placed in the hands of mankind a prime mover capable of producing at most some 40—60 HP, like the windmill introduced in the same period. Water-mill and windmill (along with the treadmill and the horsemill!) dominate technology until the end of the eighteenth century and their capacity both determines and limits the machinery, processes and products used and evolved during that period.

The fourth stage is heralded by the coming of the steam engine, which by 1850 became a prime power capable of producing more energy than the water-mill and wind-mill could ever give and capable of being erected anywhere. In our days we are on the verge of a new fifth stage, which will begin when atomic energy will be commercially available as a power resource (or rather large quantities of cheap and transportable energy in the form of electricity), the results of which can not even be guessed at the present day.

Here we have to discuss the third stage, the introduction of the water-mill in the classical world and the changes it wrought in classical technology.

Slaves and harnessed animals

During the whole of Antiquity the available prime movers were manpower and harnessed animals. Even when the more economical form of water-mills was devised about the turn of our Christian era their penetration was so slow that the situation remained practically unchanged for another four hundred years.

Manpower was always readily available in Antiquity, only in the later Roman Empire do we hear of acute shortage of labour, which incidentally contributed to the introduction of the water-mill and other machinery. The state generally disposed of concentrations of manpower in the form of statute labour or corvée for the execution of great public works or monuments. It also disposed of gangs of slaves, mostly war captives or criminals if concentration of energy was needed. Larger pieces of machinery like the large cranes used in architecture, the water-wheels draining the mines and sledges for the haulage of heavy statues were worked by gangs of workmen, slaves or animals. Often such cranes or water-wheels were worked by treadles (olivers) moved by several slaves or animals. Gangs of soldiers or slaves manned the galleys and (apart from sails) provided the motive power of ships, they built the roads and aqueducts. We must recognize that working with so large a number of "power units" already demanded efficient planning and organisation.

A few words must be said on the problem of slavery, which was of course a recognised social institution throughout Antiquity. There is, however, no basis for sweeping statements such as "slavery impeded the use and evolution of machinery and engineering". During the 3500 years covered by the term Antiquity slavery changed considerably. Both locally and regionally the aspects and consequences of slavery varied largely. Though this problem has hardly been studied in sufficient detail a few outstanding features can be formulated.

In Egypt slavery played only a minor part in economy until well in the Hellenistic period and the evolution of the crafts was in no way impeded by its existence (2). In the Near East slavery was part of the economic pattern from the beginning. Apart from the temple slaves and those belonging to the state the proportion of the unfree population in every country and at almost every time was insignificant in relation to the free population. Estates or a mining industry employing masses of slaves were non-existent. The large land-owners preferred tenants to slaves, as the latter proved expensive on the farm as well as in the shop. The basis of Near-Eastern economy was the free tenant farmer and share-cropper in agriculture and the free artisan and day labourer in the workshops. The slave had his place in the household, and always remained half free and half property. The Near East also knew the bondsman (muškennu) who worked as a kind of contract-coolie to pay off a debt or to earn a lump sum of money. He was always a paid servant and never a slave (3).

In the free Greek cities the crafts operated under a system of production in which small-shop labour was competing with household handicrafts. In the artisan shops, such as there were, free men and slaves worked side by side. The rate of pay for workmen of either status was equal, so far as our sources indicate. This Greek system of employing

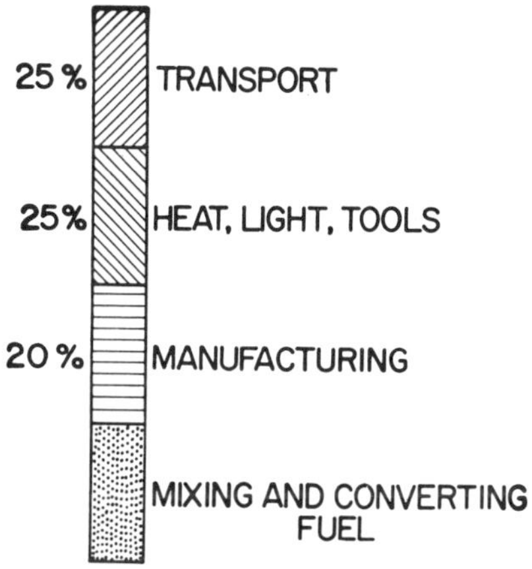

Fig. 12.
Our present use of energy

slave labour craftsmen was certainly uncommon in Italy before 350 B.C. and then it grew only slowly as the Romans conquered the world.

A considerably increase in the workshop employment of slaves occured during the second half of the third century B.C. (4). From this time onwards a nucleus of slaves was employed on the large estates (latifundia) as a dependable and permanent group to carry on the routine work of the big-farm system. However, all Latin agronomists agree that this would not be sound economically if these farms specialised say in growing wheat as they would not be fully employed throughout the year in this case. They recommend a combination of viticulture, oliveculture, vegetable growing and the raising of poultry to exploit this expensive form of labour fully. The growing of wheat and barley remained large in the hands of free farmers. Handicrafts, busines and domestic services absorbed the larger part of the slaves apart from the public works and mines run by the state. The slaves serve as a continuous body of experienced workers in the water, fire and police departments of the city of Rome.

The central period of the extension of the use of slave labour alongside with free artisans in the skilled trades in Italy (200 B.C.—100 A.D.) is roughly contemporary with the beginnings of a psychological change in the accepted Roman attitudes toward life in general and the role of the Roman state in the world. This very same period sees the beginnings and growth of manumission of skilled slaves and their

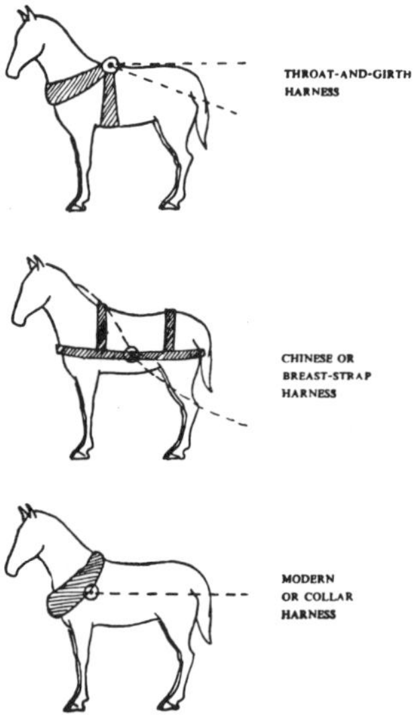

Fig. 13.
Evolution of the modern harness

growing influence both as slaves and freedmen in business and politics. The humanistic doctrines of the Stoic philosophers were generally absorbed as they became more fundamental and outspoken about the turn of our era.

Leaving aside the exonomic consequences of slavery for the moment we must turn to the question why *harnessed animals* did not largely take over the part played by human labour in Antiquity. This is largely due to the insufficient knowledge of animal anatomy which caused the ancients to use the ox-harness for donkeys, mules and horses too with disastrous effects. The ox was of course largely used for ploughing and to a certain extent to pull heavy loads and carts in the country-side.

Beyond the farm the ox was too slow to contribute to the solution of the problem of land-transport beyond a few exceptional cases, nor could it be employed in the workshops or mills in the cities.

However, in applying the same ox-harness to the donkey or horse the ancients robbed these animals of most of their natural tractive power (5). Though in Antiquity too the horse used as a pack-animal could carry about four times the load of a man like in our days, the harness then used prevented him from giving his full pull to the cart or machine. The ancient harness prevented the horse, when used as a draught-animal, to exert more than part of his available energy, for the harness did not make use of his shoulders but choked him if he pulled hard. Hence instead of pulling 15 times the load a man could move, the horse in Antiquity barely pulled four times this amount. Generally speaking therefore a horse could carry or pull just four times the weight a slave could move. The Roman agronomists, shrewd economists as they were, were soon to compare this ratio with that of the amount of food consumed by the horse and the slave, which also happens to be 4:1. Hence it would not be more economical to use the horse on the farm if plentiful manpower were available and such is their actual answer.

Usher has pointed out that we have a few data on the average tractive effort expected from teams of animals in XENOPHON's *Cyropaedia* and the *Theodosian code*, which all imply slow transport with walking animals. The wagons of antiquity were relatively lighter than ours, they are estimated at roughly one-half the effective load.

In order to be able to figure the average work to be expected in the long run under conditions that are not particularly favourable Usher quotes figures from RANKINE's *Useful Rules and Tables:*

TABLE VI

Muscular Power of a Man and of Various Animals

(Values of the greatest day's work)

1 horsepower = 33,000 foot-pounds per minute or 500 foot-pounds per second

	Pressure exerted pounds	Velocity: feet per second	Footpounds per second	Ratio
Average draught horse	120	3.6	432	1.00
Ox	120	2.4	288	0.66
Mule	60	3.6	216	0.50
Ass	30	3.6	108	0.25
Man, pumping . .	13.2	2.5	33	0.076
Man, turning winch.	18	2.5	45	0.104

The draft animal for heavy work was the ox; where speed was essential the horse was used by the ancients for light work. However, horses would be at a disadvantage if driven in small circles, e.g. working rotary mills and grinding corn. The few representations showing horses working such mills present an arrangement that is very inefficient if really constructed as indicated. Per day a horse working under modern conditions would perform ten times as much work as an average man. Mules and asses are rated as high or at least rather more than 80% of the capacity of the horse. The camel is rated at more than twice the strength of the horse, but it appears rarely in the Mediterranean even in Imperial times, even when it had become the common pack-animal of the Near East.

We also have good figures on the capacities for tractive effort as measured in terms of the useful load transported on level ground which are important to interprete the few data from classical sources.

From these figures we may say that the net effectiveness or ancient draft animals was not more than one-third of the modern expectation. This is a moderate statement and it does for instance not take into account the improvements possibly introduced by better breeding. Anyway the discrepancy between the useful work of the large animals and men was correspondingly less than now.

TABLE VII

Tractive Effort of Man and Horses

(Net loads transported horizontally)

	Net load drawn or carried, pounds	Velocity, feet per second	Load conveyed horizontally, 1 foot per second	Ratio
Horse:				
Walking with cart always loaded	1,500	3.6	5,400	100
Carrying burden:				
Walking	270	3.6	972	17.9
Man:				
Wheeling load in a two-wheeled barrow	224	1.66	373	6.9
Wheeling load in one wheeled barrow . .	132	1.66	220	4.1
Carying burden . .	90	2.5	225	

The harness devised for horses was positively inefficient because of two serious errors in its construction. The point of attachment was the back of the horse's neck, and the collar or band was set high on the throat instead of being lowered to rest on the shoulder blades. The result would be disastrous under any serious heavy tractive effort. The pressure of the collar would choke the horse, he would rise on his hind feet to relieve the pressure and would not be able to throw the weight of his body into his work as with the modern harness. (Fig. 13).

The ox may have been less inadequately harnessed and its power more effectively utilized but it was not suited to land transport due to the absence of any protection for the feet. Hence it played no important part in haulage on the road nor would it be the animal to move machinery in the larger manufacturing centres. Even its efficiency in ploughing was still less without the modern yoke.

This indiscriminate use of the "throat and girth "harness hampered the more frequent use of the domesticated animals in industry. The flour mills of ancient Rome were driven either by two men or a donkey. There was no economic incentive to prefer the draught-animal. In the early Middle Ages the "breast-strap" or "postillon" harness became known through the tribes of the steppes invading Europe, who also knew the more modern "collar" harness since the seventh century and who by the twelfth century came to use it almost exclusively (6). Between the tenth and twelfth century the "collar" harness was generally adopted in Western Europe and now the horse was allowed to exert its full tractive power. Soon the horse began to displace the ox as the motive power of ploughs and farming equipment even though it consumed more expensive food such as oats!

The same reason prevented the horse from being used economically in ancient transport. We know various data that it could not pull more than 62 kgr. and this made land transport very expensive and prohibitive in the case of mass products such as wheat. Neither did the ancients know an efficient way of yoking more animals to a waggon except by increasing the number of yokes. A much better energy output was obtained when the arrangement in line tandem was adopted in the early Middle Ages.

Finally the lack of proper horseshoes impeded the use of donkeys, mules and horses in transport and industry (7). The ancients knew metal, leather or straw "solae" (hypodemata or hipposandales) temporarily attached to the legs of horses, mules or camels when the ground was hard or slippery. The true iron horseshoe, attached to the hoofs permanently with iron nails, seems to have been adopted from the nomads of the steppes by German, Aleman, Helvetian and Suebian tribes about the second century B.C. In the northern provinces of the Roman Empire the horseshoe penetrated very slowly until its general adoption in the eighth century A.D. By that time a new type of horses-shoe with a better grip on the hoof had been evolved, which is the ancestor of the local types which we can already distinguish as early as the ninth century A.D.

The lack of proper methods of harnessing and protecting the hoofs of draught-animals practically left the ancient world with man as the only prime mover until the advent of the water-mill, the story of which we must now first trace before the factors impeding its introduction can be discussed.

The Greek or Norse mill and the water-turbine

Unfortunately the earliest history of the two first prime movers of mankind is still shrouded in mystery which only a diligent search in Akkadian, Arabic and Hebrew texts and achaeological research in the Near East will be able to dispell.

The oldest evidence of the existence of watermills is contained in a poem of Antipater of Thessalonica, who is usually believed to have lived early in the first century B.C. This poem runs (8).

> "No longer lay your hand on the millstone,o,ye
> women who turn the quern.
> Sleep long though the song of the cock announces
> the break of the morn.
> Ceres has commanded her water-nymphs to perform
> the work your arms did,
> They fling themselves onto the wheel and force
> round its axle-tree which by means of mobile rays
> rotates the mass of four concave mills."

It will be clear from this text that it does not particularly refer to an overshot water-mill as some claim but to a water-mill in general.

Next comes Strabo's information on the water-mill at Cabeira on the Lycoa in Pontus which Mithridates erected near his new palace in 65 B.C. (9). These oldest watermills were all corn-mills, for the grinding of corn was a constantly recurring burden in every ancient household and the most potent factor behind the trend to mechanize the corn-mill.

The most primitive type of mill was the Greek mill, now often called *Norse mill*, technically speaking a horizontal wheel, vertical shaft, directly coupled water-mill. A vertical shaft or axle bore a small horizontal "wheel" composed of six or eight scoops at its lower end. The shaft passes upwards through the lower millstone and it is fixed to the upper one by a cross bar or "rynd" spanning the aperture or "eye" of the stone. Hence it is also called the *vertical water-mill* (10). It requires a running stream or water for proper efficiency but does not necessarily need a millpond and all other precautions to regulate the water supply the later types demand.

We do not know the origin of this simple watermill but historically it was always bound to mountain regions, and very often has a mill-race and shute to improve its efficiency. It is essentially an element of peasant culture. Because of its low efficiency it is not suitable for commercial production of flour but each mill serves the limited need of a private farmer. The stones of small diameter move slowly, they practically always turn once per revolution of the water-wheel itself. The entire mill is very small and usually incapable of grinding more than a sack of corn at a time. Some have claimed it as a product of Greek mechanical science, but its country of origin probably is the mountain region of the Near East (11) whence it spread to the west and the east. We should note here that they were never found in the river valleys of Egypt and Mesopotamia probably because of the great fall and rise of the level of the big rivers. None of the Hellenistic papyri from these regions mention them. They were unable to use anything but a small and constant mass of water moving at high velocity common in hilly districts, the river-valleys could offer only large masses of water at a low velocity.

However they were quite well-known to VITRUV (12) and to LUCRETIUS who significantly says that there may be "another current beneath, to rotate the globe (of the sky) reversely, as we see streams moving round wheels with their scoops" (13). In the first century A.D. PLINY (14) points out that "in the greater part of Italy a roughened pestle is used, and wheels also that water turns round as it flows along; and so they mill". This probably refers to northern Italy and possibly Apulia and Sicily, where VITRUV places them too.

They spread to China in the third and fourth centuries A.D. and reach Ireland about the same time. Recent excavations at Bolle (Denmark) (15) tend to prove that these mills were known and built in Denmark by the birth of Christ, their number increasing rapidly in the following centuries. The two Early Iron Age mills were provided with a dam and mill-races. They were very popular until the Late Middle Ages. HENZER found them on the Garonne (France) as late as 1588. Now the Norse mill is still used in the Orkney and Faroe Islands, Norway and Roumania. They survived in some isolated places like Mount Athos, the Lebanon and Mount Carmel (16) and they are also used in many parts of Central Asia, but they have now disappeared from the mountainous regions of Wales, Eire and Scotland.

Though the Norse mill provides too little power to give anything but extremely slow movement to the mill-stones and therefore guarantees only a small output of flour sufficient for home consumption it is important as the precursor of the modern water turbine. Earlier forms of these water-turbines seem to go back beyond the Middle Ages, when

Fig. 14.
A Norse mill in a German mss of 1430

a kind of Pelton-wheel, built in 1430 near the monastery of St. Georgenberg in the Inn-valley, was said to have been inspired by the Pope at the time of the Council of Basle. In the regions of Province and Dauphiné they continued in use upto the eighteenth century, and likewise in Umbria and other regions of Italy (17).

More modern forms of waterturbines were introduced about 1775, however, their proper calculation and construction started only with the work of Fourneyron (1823) and his generation.

The Greek mill, though mechanizing corn-milling for the private farmer would have had little effect had it not inspired the Roman engineers of the first century B.C. to construct the more efficient *horizontal or Vitruvian mill.* Technically speaking this was a vertical wheel, horizontal shaft, geared water mill. Because of his knowledge of gear-wheels and all the other achievements of Hellenistic mechanical skill the inventor transformed the old Greek mill into a much more useful piece of machinery. The water-wheel turned by the impact of water was placed in a vertical position. In order to enable the horizontal shaft to move the millstones it had to be geared to the vertical shaft of these stones. In Roman water-mills the millstones usually made five revolutions to one of the wheel. The much greater efficiency of this new design outbalanced its more complicated construction. Two sub-types should be distinguished, each with its own advantages and limitations.

The *overshot-wheel* is a water mill the water of which impounds on the top of the wheel on the scoops or buckets built into the circumference. It therefore needs a well-directed and regulated water supply, mostly tapped from rivers and wells and collected in a mill-pond, then delivered

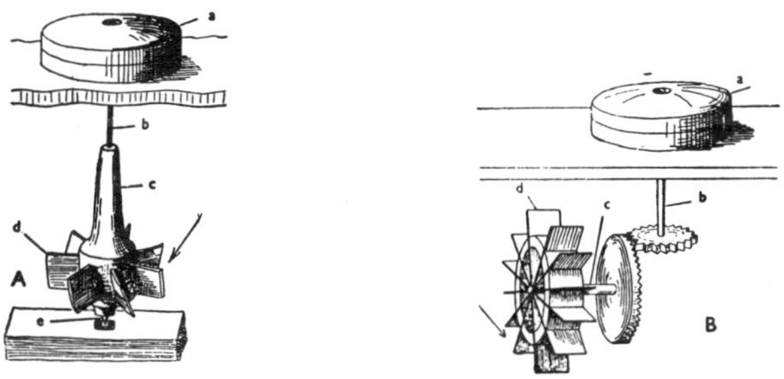

Fig. 15.
The Norse mill and the Vitruvian water-mill

through a sluice to a millrace and chute properly set for the correct impact of the water on the wheel. The water of sluggish rivers or rivers of variable volume can thus be put to advantages, the river level being raised by a weir or dam to spill its overflow on part of its waters into the mill-pond. Aqueducts are equally suitable suppliers of water.

This more elaborate but most efficient type of water-mill was not too common in classical Antiquity. The earliest example comes from the agora of Athens and dates of the third quarter of the fifth century

A.D. Even in the Arab world it did not predominate for AL-QAZWINI considers worth mentioning (18) that Al-Adri (from Abdera) reports an overshot wheel near Waluta (Majorca) fed by a shute 80 cm in diameter and 350 cm in length. It was built because "if the water falls the (common) mills cease to turn" (19).

This goes to show that the type common to Antiquity and even many centuries afterwards was the less efficient *undershot-wheel*, the type described by VITRUVIUS and probably the first to be designed and built. It works efficiently only in swiftly flowing rivers and fairly constant volumes of water, and does not necessarily need a dam, a millpond and millrace. In fact the combination of dams (or weirs) and mills becomes general only in the thirteenth century, the earlier mills being fed by

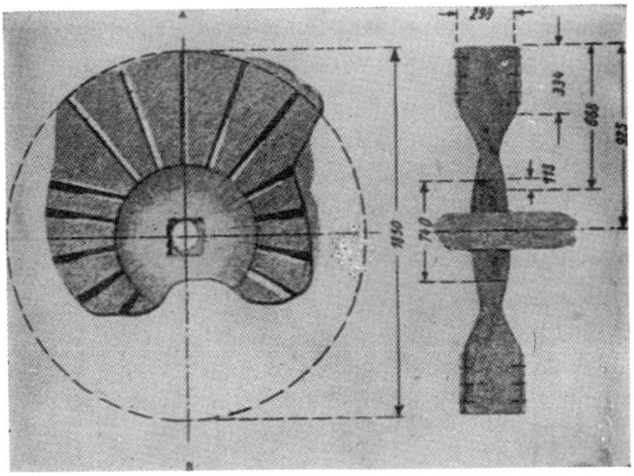

Fig. 16.
Construction of the Venafro mill (After Reindl)

aqueducts if need be. The undershot wheel is less efficient because no use is made of the difference in level between water supply and drain as in the case of the overshot wheel and also because like in the case of the Greek mill the full impact of the current can only be applied if a shute is used. Their inventor is unknown but seems to belong to the first century B.C.

The earliest text call the water-mill simply "mola" or "cerealia saxa" (20). VITRUV calls it "hydramula" or "hydraletes" like STRABO and PLINY (21). The latter term gradually clings specifically to the new type of water mill. The actual remains of three of these early mills were discovered, all built during the Roman Empire.

At Venafro (the ancient Venafrum) on the Tuliverno near Naples the mill-race and remains of an aqueduct were discovered together with the imprints of a water-mill in tufa-sediments for the millstones were covered by no less than 3 m of sediments (22). The heavy nave (diameter 74 cm = 10 Roman digits!) carried 18 spokes, formed like a spade. Two parallel rims joined the spokes, thus forming a wheel of a height of 25 Roman digits. This undershot wheel fed by an aqueduct had an output of 3 HP and the millstones performing 46 revolutions

Fig. 17.
Plan of the Arles flour-mill (After Benoit)

per minute could grind 150 kgrs. of corn per hour. If we compare this with the usual 7 kgrs. an hour ground by two slaves in hand-querns or the 0.4—0.5 HP developed in the donkey-mill we see clearly that this new Vitruvian mill meant a revolution in corn-grinding, once it was generally applied. The energy-output of this device was much higher than anything driven by man or beast, it was able to produce an amount of concentrated energy beyond any other power resource of Antiquity. In other words, once it was realised that the water-mill could not only grind corn but also move other machinery larger machines could now be built, larger pieces of material worked and new things created which upto now had been beyond realisation.

This is clear when we turn to the second example, the flour-factory of Barbegal, six miles from Arles. Alongside the aqueduct of Saint

Rémy conducting water to Arles and built in the days of Agrippa we find a second one collecting the waters of the Arcoule, the valley of Les Baux and the well of Manville. This aqueduct of Les Baux has a duct (specus) 2 m wide and 5.6 m deep and it is built of bricks and concrete (containing pot-sherds), hence it is younger than the reign of Hadrian. At Barbegal (23) the specus is doubled and on a slope of 30° with a fall in level of 18.60 m two sets of eight undershot-wheels are built in these conduits. These wheels had a width of 70 cm and a height of 220 cm. On the iron axle-tree wooden gear-wheels were fixed with lead. The gears were built in chambers under the millstones.

Fig. 18.
Reconstruction of an overshot mill which was operating in the fifth century A.D. in the Athenian market-place
(After Stork and Teague)

Each wheel turned only one pair of stones and not two as in Athens. Thence the stones could be made heavier and more efficient. The lower stone was 45 cm thick and its diameter was 90 cm. The upper stone had an "infundibulum" or funnel feeding the cereals to the milling surface.

Comparison with older types of corn-mills has revealed that the capacity was 150—200 kgrs of corn per set of stones, or a total capacity of 2400—3200 kgrs per hour. Even in a ten-hour day this would mean 28 Tons of flour. This would be sufficient to feed a population of 80.000 but Roman Arles counted hardly more than 10.000 inhabitants in the third century A.D.

The coins and pottery found near this flour-factory date of the third and fourth century. Arles was an important harbour for the corn supply of Rome and the origins of this factory may go back to the days of Marc Aurel early in the second century when "the procurator of the annona of Narbonensis" settles at Arles and for some time even looked after the harvests of Liguria. Benoit supposed that this flour-mill was built by Q. Candidius Baningus, the famous local engineer who according to an inscription "was clever like none other and none surpassed him in the construction of machines (organa) and the building of water-conduits". This suggestion can not, however, be substantiated by any further proof.

In the present state the mill seems to have been constructed in the days of Theodosius who resided at Arles from 308 to 316 and established a mint there in 313 A.D. At that date Arles also had important "horrea" for the storage of corn and the flour-mill may have served for local consumption and that of the army of Narbonensis. We have literary evidence of a similar flour mill at Tournus (Burgundy) which ground the corn of the Saône valley and which was situated in the little village of Préty (Pistracum) near the well of Bouat, feeding the army of northern Gaul.

The third wheel discovered by the archaeologists was an overshot-wheel in the Athenian Agora (24) built against the Valerian wall. The diameter of the overshot wheel turning two sets of millstones was 324 cm. The wooden axle-tree was made of one single beam, some 350 cm long, and turned in wooden bearings. The grain seems to have been dried before milling. The calculated power output was probably double what was actually delivered.

Archaeological evidence places the building of this mill in the long reign of Leo I (457—474). The mill was probably destroyed by fire during the reign of Justinus II (565—578), may be during the great Slav invasion of 577. This mill with its mill-race with a single trap implies a copious and steady stream. The water supply from Hymettos and Pentelikon brought in by the Peisistratids was relatively modest. However, excavations of the Agora have shown a series of springs used from the earliest times and probably originating from the north slope of the Areopagus. Even at our late date they were still sufficient to supply a Roman bath and our mill with the necessary amount of water.

The city of Athens that began to expand in the third century (after the building of Valerian's wall in the second century A.D.) no longer had the big fountainhouse and it was little more than a small university town. That fifty years later a mill could be built only a stone's throw away and fed from the public water supply is a sure indication of rapid decline. After Justinian's edict of 529 the doors of the University closed and Athens became a village, but the mill survived another fifty years.

The earliest picture of a water-mill is an early fifth century mozaic uncovered in the Grèat Palace of Byzantium (25). It shows an undershot wheel of the Vitruvian type and looks similar to the one found at Venafrum.

The introduction of the corn-mill

The rarity of the archaeological sources and the absence of disputes over mill-rights and yields of flour (both so common in medieval texts) form an impressive coincidence. When therefore in the fourth century A.D. the data show a rapid increase in the number of the water-mills, when they come into use on the confines of the Roman Empire as testified by Ausonius (*Mosella*, 361—364) (26) (who refers to water-mills for cutting and polishing marble on the Ruwer (an effluent of the Mosel) in 379 A.D. which passage we will discus later on) when we find an epitaph at Sardes in the East (27) mentioning a "manganeiros hydraleta" and tradition says that the Hellenized Persian Metrodorus introduced the water-mill into India during the reign of Constantine in the early fourth century, we should look for the reasons why the use of these mills caught only then after remaining comparatively unimportant for so many centuries.

These reasons can be best understood if we focus our attention on the production of flour in the city of Rome itself during the Empire. Sextus tells us in his *Topography of Rome* written about 200 A.D. that the city had 300 mills but does not specially mention any water-driven mills. Thus during the Empire corn-grinding had become a specialised trade (like that of the baker (pistor) by 174 B.C.) centralizing an operation formerly part of the every-day routine of the Roman household and relieving the baker who had formerly produced both flour and bread of part of his task. The centre of these bakeries and corn-mills was the Janiculum. The type of mill used was the horse- or donkey-mill. When the emperor Caligula confiscated the horses in the bakeries of Rome, a bread famine resulted.

Some of these mills were even man-driven, slaves or unskilled poor supplying the necessary labour. During the early Empire the problem of the unemployed was still rampant as is clear from many stories about labour-saving inventions which the various emperors are said to have refused or destroyed in order to give these unemployed a means of earning their living or to provide prisoners with work. Constantine the

Great, who already had to cope with a short supply of labour in his days, added the flour-mills to the places of penal servitude. Hence about 300 A.D. part of the mills were still worked by criminals, most, however, were horse- or donkey-driven.

An acute shortage of labour in the fourth century helped considerably towards the use of the water-mill for grinding corn. With this labour shortage in mind PALLADIUS (28) recommended the use of water-mills to ease the burden of men and animals. He mentions corn-mills driven by the water of public baths and aqueducts. We know indeed that the overflow of the Aqua Traiana was used for manufactures housed on the Janiculum on the north bank of the Tiber and that water-mills were constructed in the baths of Caracalla dedicated in 216 A.D. (29).

Now water-mills begin to invade the Janiculum despite the opposition of vested interests and the owners of the old type of horse-mills. The new mills often diverted water from the old aqueduct of Trajan and we have an edict of 398 A.D. in which Honorius and Arcadius forbid this misuse of water for the cluster of water-mills on the Janiculum (30). It seems that they were mainly fed by the waters collected in the neighbourhood of the "lacus Sabatinus". PRUDENTIUS (*contra Symm.* II, 949—950) (31) mentions them and PROCOPIUS tells us about the Janiculum: "This is a region across the Tiber where rise several tolerable hills, and where now, as in former times, are erected all the mills, a large body of water being conveyed from the top of the hills by timber structures causing the water to fall down the slope with considerable force." (32)

Again and again we find edicts like that of Zeno of 485 confirmed again by Justianus in 538 A.D. (33) and that Theodoric (about 500 A.D.) (34) forbidding the millers to divert water from the aqueducts to their mills showing the increase of their number right into the sixth century. Further laws are contained in the *Digests* of Justinian (35) and in the later Visigoth laws which say: "Of the fracture of mills and water-sluices. Any one violently injuring a mill shall repair the injury within thirty days; and the same with regards to pools and sluices attached to mills." (36)

The evolution of the water-driven corn-mills gave rise to a new class of craftsmen, the "molitores" or "molendarii", the millers generally owing the mill, grinding corn and sometimes even preparing malt or baking bread. The latter craft, however, generally remained in the hands of the bakers, whose guild, the Collegium Pistorum had been founded by the emperor Trajan about 110 A.D.

Obstacles retarding the introduction of the water-mill

It is obvious from the above data, that the Roman engineers of the first century B.C. had transformed the primitive Norse mill into a real prime-mover. How is it that its introduction was delayed until well in the third century A.D. and even then was restricted to a few centres only?

Several factors combined in delayed this introduction and one springs directly from the physical geography of the Mediterranean basin itself. Most rivers in this centre of the Roman Empire carry widely varying quantities of water in different seasons and this greatly hampers the introduction of undershot-water-mills constructed on a masonry foundation. During a large part of the year the water-supply would be insufficiently to turn the wheels. Hence the building of water-mills was restricted to those few spots in this area where a constant and liberal water-supply was available from river or aqueduct and where there was at the same time a heavy demand for flour and a lack of suitable labour or where wages were relatively high. It is typical that all the water-mills excavated (those of Venafro, Barbegal, Tournus, Rome and Athens) were fed artificially by an aqueduct in order to have a constant watersupply. This heavy investment of capital would tend to limit the introduction of the water-mill to the great centres of the Roman world such as Rome, Athens or Byzantium. Only in a few cases would such military centres as Barbegal or Tournus qualify by their importance for the erection of "flour-factories". Labour too was plentiful and cheap until the end of the third century, precisely the period when the donkey-mills or slave-driven mills of Rome were gradually ousted by the water-mill.

In Western Europe the geographical conditions were different. There were many mountain streams and brooks whose constant water-supply favoured the introduction of "a mill on every manor thar had a stream to turn it" during the Middle Ages.

Still this does not fully explain why the Roman engineers applied the water-mill to the production of flour only and why they never tried to introduce it as a prime-mover except for the few instances mentioned at a very late date. It may be true that the Roman engineers were not particularly ingenious and that most of their knowledge consisted of a practical application of the inventions of the Hellenistic engineers of the East (37), but they did possess a body of practical experience to enable them to build machines moved by water-wheels.

Their knowledge of the principles of the lever, the winch, the wedge, the screw and the pulley, of the more important principles of hydro- and aero-dynamics enabled them to construct machinery to be moved by such water-wheels as they did in the case of the corn-mill. In fact their practical knowledge of machinery was not significantly surpassed until the eighteenth century (38).

We have sufficient examples showing that this knowlege could be translated into practical results. Indeed where public opinion demanded it and the state applied sufficient pressure and supplied the money mechanisation could go very far indeed in Antiquity. This is the case in the field of warfare where from Hellenistic times onwards mechanical missile weapons based on the bow or sling were developed and built in large quantities. Tests on such machines have proved their efficiency and the number of such ballistae and catapultae introduced in the Roman army points to a mechanisation that was not surpassed until far in the nineteenth century.

In the field of achitecture large cranes were constructed and driven by treadles. Suetonius records that Vespasian gave "to a mechanical engineer, who promised to transport some heavy columns to the Capitol at small expense... no mean reward for his invention, but refused to make use of it saying: "You must let me feed my poor commons." (39). Manual labour was actually created and the corn dole continued so long because it served as part of the wages given to the indigent employed on public works. Large water-wheels draining the Spanish mines continued to be worked by treadles or treadwheels.

Except for state services and public works there was no demand for mechanisation. All arts and crafts were left to themselves to work out the best trend in their own trade. Many crafts like spinning and weaving were still mainly household occupations and hence show little progress. Except in certain classes of pottery the classical potter achieved little progress, he produced excellent thin-walled pottery but failed to produce tin- or salt-glazes which had to be fired at temperatures beyond his primitive kilns (40). The lack of stimulants to industralize left ancient technology practically stagnant during the Roman Empire. In a few instances e.g. in pottery, shoemaking and jewelry we hear of some specialisation tending to regional mass-production of certain goods (41). In most case the manpower available tended to counteract all attempts of the engineers to mechanize trades and to introduce machinery. Their efforts were only too often wasted on showpieces (thaumata, mirabilia) meant to impress the masses at public festivals.

Not only did ancient technology lack any trend towards economic efficiency, above all the fertile interplay between *science and technology* was almost totally absent. The ancient engineer was very much aware of this necessity for he says: "The mechanicians of Heron's school say that mechanics can be divided into a theoretical and a manual part; the theoretical part is composed of geometry, arithmetic, construction, astronomy and physics; the manual of work in metals, construction work, carpentering and the art of painting, and the manual execution of these things. The man who has been trained from his youth in the aforesaid arts and in addition has a versatile mind, will be, they say, the best builder and inventor of mechanical devices. When it is not possible for the same man to excell in so many academic studies and at the same time to learn the aforesaid crafts, they advise one who wishes to undertake "mechanical work" to use such crafts as he already possesses in the tasks to be performed in each particular case." (42) Only in the fourteenth century did these superior craftsmen succeed in making themselves heard, in Antiquity science took no lasting notice of them (43).

The ancients had no word for "scientist", they called him a "philosopher". They might be "physiologoi", that is they would investigate "the natural growth of things" (44). They observed nature and took many of their images from the crafts of the Greek towns. Mechanistic thought-patterns are used almost exclusively to explain certain natural phenomena, but mostly their observations serve to adstruct their philosophical theories (45). They have little or no place for manual labour or anything connected with technology and engineering. Such important scientists as physicians to them were "technitae", that is "craftsmen", their work included manual operations and hence they were just skilful craftsmen. Their study of Nature formed part of the pursuit of peaceful wisdom and happiness to which each free citizen of the classical world would devote his "otium" or free time. Anything like mechanical arts or "neg-otium" was definitely inferior; by excercizing their trade the banausoi (craftsmen) "sitting indoors like women" killed the spirit (46). PLUTARCH calls the machines Archimedes made when the Romans besieged Syracuse "the byproducts of a jesting geometry" and says that "the construction of engines and in general every trade that is exercised for its practical value is lowly and base". In this very same spirit neither Plato nor Aristotle wanted the citizen to be a craftsmen.

This typical Greek contrast between the liberal arts and the "artes mechanicae" precluded all efficient cooperation between science and technology. Even the effect of trade knowledge on science is extremely small and doubtful (47). The classical scientist at his best had a clear apprehension of scientific method, he was aware of the need of deduction and verification and conducted experiments. De Santillana pointed out that Plato and his school moved away from the more scientific attitude of the earlier Ionian scientists and replaced their physics by an astro-theological cosmos which discouraged experimenting, though such experiments are still conducted now and then after the sixth century B.C. (48). But he was too empirical in the naive sense and little did he analyse the difficulties involved in adequate observation. His results might be of value to his philosophical theories, they yielded no data that could help the engineer and the craftsmen who were left to travel the hard road of trial and error alone. The Baconian notions of "utility" and "progress" did not belong to his world. Bacon himself was clearly aware of this when he said: "Great technical discoveries were more ancient than philosophy and the intellectual arts: so that, to speak the truth, when contemplation and doctrinal science began, the discovery of useful works ceased." (49)

Nor did *ancient economy* stimulate the harnessing of the forces of nature by prime mover in order to produce great quantities of cheap goods for the masses. Some goods were indeed produced on a large scale, cereals, wine and olive-oil and to a certain extent pottery too. In the case of the first three products this mass-production resulted indeed in the development of better presses, roller-mills and corn-mills but the available men-power at low wages put a halt to further mechanisation. Simple pottery, the packing material par excellence of Antiquity, was still produced in small workshops. There are but few signs of specialisation except in the case of some more expensive types of vessels connected with the storage and drinking of wine and of some better wares like Arretine and other types of pottery.

Nowhere in Antiquity do we find a drive for industrialisation, demanding more concentrated output of energy by prime movers and leading to specialisation, mechanisation and standardisation. The "fabrica" of the ancients was the workshop which dealt with products now classed as heavy industry, that is mostly a smithy. The larger grouping of craftsmen in an "ergasterion" or "officina" was hardly a factory in our sense. The Athenian ergasteria with 10—30 craftsmen were not the result of heavy capital investment for the purchase of machinery. They housed a series of craftsmen of the same trade and corresponded rather with the "ateliers" or "manufactures" of the eighteenth century. The net profits of these workshops were not reinvested to increase their output, they were consumed by the craftsmen themselves, whose property in tools and private possessions was very small as we read from papyri of Hellenistic Egypt (50).

The wealth of the classical world was concentrated in the hands of a small group of citizens, notably during the Roman Empire. Their attitude towards the value of applied science and the conquest of nature for the benefit of all mankind has been sketched above. The masses lived on the verge of indigence except for a few prosperous periods such as the early days of Hellenism and the first two centuries of Roman domination. This concentrated purchasing power demanded certain luxury goods, which were imported, mainly from the East or which could be manufactured in relatively small shops. Many of our larger industrial products such as textiles and food were produced by the household, not even by craftsmen. Nowhere except in towns near larger centres did production rise to anything of the size of an export industry. Even the larger workshops of Capua and the Campania worked mainly for Rome and not for consumption in outlying districts of the Empire. Production was mostly consumed locally in each area and there were no great demands on land or sea-transport for long-distance trade.

This agglomeration of production areas, in each of which autarky reigned, suffered little of the sharp economic crises of our modern world, over-production was practically non-existent. Mass-production of goods would in fact have meant over-production in the ancient world, which invested its capital in slaves and land. No large sums were available for the development and construction of machinery to replace human or animal power, cheap enough in a world which produced the bare necessities of life for the masses and a few luxuries for the few. Applied science had already completed its task, so the ancients thought and this may have been one of the causes of the failure of the Roman Empire as it was of the Hellenistic world of the East (51).

In this world which did not know the employer and labour class of our days modern economic factors and urges did not apply. Only in the larger centres did the mechanical production of flour, one of the necessities of the masses, become urgent by the fourth century when the expansion of the Roman Empire had long since stopped and labour became more and more scarce, even in the form of slaves. By then the Roman authors begin to complain bitterly of the neglect of applied science and, not without reason, to extoll the inventiveness of the barbarians (52). "In fact this is a quality which we see granted without respect of persons; for although the barbarian peoples derive no power from eloquence and no illustrious rank from office, yet they are by no means considered strangers to mechanical inventiveness, where nature comes to their assistence", says the author of the *De Rebus Bellicis* about 370 A.D. (53), whose inventions are specially designed to save manpower in the army. Once this wide-spread social demand for cheaper bread existed the vested interests in slave- or donkey-driven mills could easily be overcome by introducing the mechanical device already known for centuries.

By the time the water-wheel is introduced as a prime mover a spiritual revolution had swept away the classical views on the degradation of manual labour. Slavery had always depressed the social and economic conditions of the free craftsmen, kept their wages low and subjected them to be contempt of the intellectual classes. Gradually the latter could not refrain from being engaged in industrial production, whether on their estates or in the workshops of the large cities. CICERO, who is very outspoken on the degradation of manual labour, respectfully winks at "Big Business" and esteems a "mandatum" in running and financing slave-operated workshops correct for a free citizen of Rome. The guilds of craftsmen were never wholly without influence in ancient Greece and Rome and by the second century A.D., swelled by an increasing number of freedmen, became stronger powers in the state. These guilds were never labour unions, but rather religious and social brotherhoods who sometimes venture to press for economic improvements but they are never involved in modern labour policies. The ancient craftsmen took a great pride in their association with their patron god, deriving their descent from them, and honouring him with offerings and festivals.

The change came in the later Roman Empire when slavery was already on its ebb. The questionable blessing of cheap slave labour combined with the restricted demands had reduced the wages of the free artisans, mechanical inventiveness and the organisation of efficient methods of production of cheap goods for all. The change came through a new *attitude towards the poor*. The ruthless suppression of the weak in classical Antiquity (notably between 200 B.C. and 100 A.D.) showed its worst face in the treatment of slaves, which in its turn led to frightful revolutions, not primarily based on social programmes, but on the elemental will to survive, to escape from captivity and to return to the homeland.

The poor were regarded as victims of the gods in the Ancient Near East and recommended by religion to the sympathy of their wealthier brethern. Only the Old Testament rose to the moral concept of the inherent brotherhood of man (54) and the denial of the right to own a man to a man in perpetuity. The bondsman was the only form of slavery allowed in ancient Palestine. In the classical world several schools of philosophers, notably the Stoics preached this brotherhood of man. They tried to instill "philanthropia" and "humanitas" into the hearts of their fellow countrymen, who should no longer treat their poor and weak brethern as chattel but as human beings. The advent of Christianity introduced a more basic concept, for which the new term "caritas" (charity) was coined (55), which was more than just friendliness and hospitality.

Christianity not only changed the attitude of the citizen towards the poor and the slave, it radically attacked the classical views and extolled the value of manual labour. By proclaiming and raising the dignity of work slavery was condemned to disappear in the long run and the craftsmen was again to take an honourable place in society. The door was opened to the view that nature should be used to serve mankind, that its forces should be captured and trained to ease his task and his life (56).

Moreover the ancients had always believed in an animated nature. To most of them it was a work of art of beautiful suitability in which Form had not always completely subjected Matter. "Matter tries to escape the geometrical forms which the logos tries to impose on it" says PLUTARCH. Others such as the atomists considered that nature had achieved its structure by boundless waste. But none of them would have accepted the "Newtonian world-machine" created by God and left to run its course, the concept so dear to the eighteenth century scientist. Even DEMOCRITUS, EPICURUS or LUCRETIUS would have shuddered before such blasphemies. In reality the ancients always based science on animism. Even THALES says that "all things are full of gods" and that "the magnet proves to have a soul since it can move iron". This animism is still and essential element in the final philosophy of Antiquity, Neoplatonism. The poet's "waternymphs moving the axle of the water-wheel" were quite real to the masses of Antiquity and not just an image. This ancient world would not dream of man harnessing these supernatural powers. Christianity did away with this animated nature by one stroke and by denuding it of its supernatural halo opened the door to a rational use of the forces of nature. The belief in these possibilities and the demand to realize them grew as the ancient world was transformed into a basically Christian one. The last obstacles for the introduction of the prime mover had fallen, when by the fourth century A.D. the Roman Empire became officially a Christian one.

The paddle wheel and the floating mill

The invention of the paddle-wheel propelled ship was inspired by the water-mill. Its principle reverses the idea of the water-mill by turning the axle to move a water-wheel, which can then propell a ship through a river or the waters of the sea. The probable inventor was the

Fig. 19.
The floating mill

unknown Latin-speaking author of the *De Rebus Bellicis* composed between 366 and 375 A.D. (57). His warship or "liburna" has three sets of oxen each turning a capstan rotating two paddlewheels on the outside of the ship. He can claim to have been the first man imagining the possibility of propelling a ship without the use of oars or sails. His invention of the paddle wheel preceeds by several centuries the Chinese claim of having used human-moved paddle-wheels in a naval battle of the twelfth century.

How completely the water-mills have displaced the older types within 150 years is evident from the fact that the Goths, when besieging

Rome in 537 A.D., can attempt to starve Rome by cutting the aqueducts. This incidentally promoted the invention of the *floating mill* by the general Belisarius and his engineers for PROCOPIUS tells us (58): "When the water was cut off and the mills stopped, and cattle could not grind, the city was deprived of food, and provision could hardly be found for the horses. But Belisarius, an ingenious man, devised a remedy for the distress. Below the bridge across the Tiber which arches to the walls of the Janiculum, he extended ropes, well-fastened across the river from bank to bank. To these he affixed two boats of equal size, two feet apart, at a spot where the current flowed with the greatest velocity under the arches; and placing mill-stones in one of the boats, he suspended machines by which they turned in the water-space between them. He also contrived, at certain intervals on the river, other machines of the like kind, and these being put into motion by the force of the water, drove as many mills as were necessary to grind food for the city."

This floating mill was to spread east and west in the centuries to come. In the tenth century floating paper mills were found on the Tigris near Bagdad, and we hear of "boat-mills" near Venice. These "molendium navale" are often referred to as "mills of Babylon", probably this is an echo of what the crusaders saw in the Near East. In the Seine floating mills were built under the Grand Pont at Paris during the reign of Louis VII (1137—1180) and destroyed with this bridge in 1296. They feature on the Garonne (1290) and the Loire near Orleans (1306). Attempts to build them on the Thames failed twice (in 1525 and in the eighteenth century). They occured in Central and North-Western Europe upto the nineteenth century and were still fairly common a hundred years ago on the Tiber, on the Danube and on several rivers in the Balkans, a few survived in Italy and other regions upto our time.

THE INTRODUCTION AND SPECIALISATION OF THE WATER-MILL

By the fourth century men had discovered the real importance of the water-mill. In the same way as many centuries later the steam-engine grew to be a prime mover starting as a steam-pump used in mines, thus we see the gradual recognition of the water-mill as a prime mover, as a power resource rather than as a mechanised corn-mill. We hear about early mills driven by the tributaries of the Moselle. Venantius Fortunatus (c. 600 A.D.), who knew the area well, describes them as water-driven flour mills (59), but the earlier poet Ausonius (c. 400 A.D.) tells us that "renowned is Celbis for glorious fish, and that other, as he turns his mill-stones in furious revolutions and drives the shrieking saws through smooth blocks of marble, hears from either bank a ceaseless din" (in Hugh G. Evelyn White's translation in the Loeb edition) (60). However, Lynn White has expressed his doubts about the authenticity of this passage in Ausonius and hence we believe that not too much value should be attached to it. Bélidor mentions a number of very ancient saw-mills and we know now that water- and wind-power were certainly quickly adopted during the early Middle Ages in Normandy, Champagne, Dauphiné and Languedoc (61). The

Fig. 20.
de Honnecourt's drawing of a saw-mill (1250)

reports that the ceiling of the banquet hall in Nero's palace was rotated by a water-mill may be wrong (Suetonius, *Nero*, 31.2) (62), but authors certainly began to suggest timidly all kinds of application of water power. It was stimulated by the growing conviction of the Christian community that the use of human or animal power should be avoided if Nature could be made to do the work.

The troubled political and social times following the desintegration of the Roman Empire were hardly favourable to a rapid spread to the water-mill. Still we see the number increasing steadily in Western Europe. The tenth to twelfth centuries form the climax of the introduction of this new prime mover (63).

Focussing our attention on France we hear of water-mills near Dyon and Geneva about 580 in the *History* of Geoffrey of Tours. The *Vita Petrum* (64) mentions one near Angers, some six water-mills seem to have existed about 600 A.D. During the Frankish Empire

Fig. 21
Goulande Mill, Jersey (1274)

they were sufficiently important to be protected in the Salic laws (65). The eleventh and twelfth centuries show the rapid progress of the medieval industrial revolution, a strong heave of mechanisation in different industries followed by a period of stability until the fifteenth century. Though we have many references to water-mills in the Capitulare de villis, we have no total figure for France about 1100. The rapid growth can be illustrated by some local figures. On the banks of the Robec, which joins the Seine at Rouen there were 2 mills in the tenth, 5 in the twelfth, ten in the thirteenth century and 12 about 1300. In the district of Forez there was one early in the twelfth but nearly 80 in the thirteenth century. In the district of Aube 14 of the eleventh grew to 60 in the twelfth and nearly 200 in the early thirteenth century. The town of Troyes used the Seine and the Meldançon to build eleven water-mills in the period 1157—1191. By 1493 the district had 20 corn-mills, 14 paper-mills, 2 tanning mills, 4 fulling mills and 1 cloth-mill, that is a total of 41.

At the same time there is a growing specialisation of mills. The industries that could use the circular motion like the corn-mill were water-lifting and irrigation wheels (abbey of St. Bertin, 1095—1123), oil-mills (eleventh century in the Graisivaudan area), mills grinding pigments (Péronne, 1376), malt-mills (Béthune, 1138), wood-turning mills (Vizille, 1347) and grinding mills for cutlery (Forez district, 1257). Other mills had to use a variety of gear-and cog-wheels to move their machinery, e.g. fulling-mills (Dauphiné, 1050; Forez, 1066; Champagne, 1101), tanning mills (Graisivaudan, eleventh century; Issondin, 1116), hemp-mills (Graisivaudan, 1040), iron mills (1116), saw-mills (thirteenth century) (illustrated in VILLARS DE HONNECOURTS sketchbook of about 1250 A.D.) (fig. 20), papermills (thirteenth century). In the south of France there is an intimate connection between the irrigation canals and the establishment of water-mills. After the building of the Vaucluse canal (1101) licences were given for the building of water- and wind-mills, grinding corn, pressing olives and fulling cloth. The large monasteries were particularly interested in the building of water-mills. In 1020 the abbey of St. Victor near Marseilles possessed a water-mill on the Jaret near Huveaune, and a branch of the Charente turned the mills belonging to the abbey of Saintes. In fact, it seems that the order of Cistercian monks, those undefatigable reclaimers of barren and waste lands, was prominent in the diffusion of the water-mill. Their abbey in southern Champagne had fulling-, paper- and hammer-mills together with water-wheels in the thirteenth century. They were particularly involved in the building of water-driven hammer forges and they appear as pioneers in iron metallurgy with the help of water-mills not only in France but also in Germany, Denmark and Great Britain. Out of 30 French documents of the twelfth century concerned with hammer forges and iron metallurgy 25 were drawn up by Cistercian monks and producers. Some of these abbeys were water-driven workshops combining very different crafts under one roof.

Tide-mills, though rare, appear as early as 1125—1133 near the mouth of the Adour. A century later they are signalled on the east coast of northern Italy near Venice. We possess a drawing (1438) of the tidal-mills by the famous engineer JACOPO MARIANO. Dover harbour had an early one dating from before the Conquest (1066). However, most of them date of the eighteenth century and they never played any important part in industry.

We observe the same spectacular rise of the number of water-mills in Great Britain together with the increase of population, the rise of cities and the growing mechanisation of several crafts. The earliest reliable allusion to a corn-mill in England occurs for the year 762 in a charter granted by Ethelbert of Kent to the owners of monastic mill situated east of Dover. Gradually horse-mills and water-mills come to concentrate corngrinding, without, however, completely displacing

Fig. 22.
Medieval water-mills under a bridge

the household quern. By the tenth century the water-mill has invaded Ireland. A century later the *Domesday Book* (1086) reveals the existence of no less than 5624 water-mills in 3000 communities, south of the Trent and the Severn. Very few were then located in Cornwall (66). Some of these mills may have been of the Norse type, though this was typical of the Celtic and Norse settlements in the hills. The horizontal wheel won out on the long run.

The water-mill altered the functions of the housewife and introduced the miller. It was a source of profit to those who erected them upon their holdings. A mill became indeed the unfailing mark of a manor that had a stream to turn it, though only one third of the manors were the sites of mills, which occur in clusters. The greatest density occurs around the central highland and in the east facing the continent. This underlines the radial distribution of water-power utilization. The ratio is about one mill on every 50 households, being the lowest in the west and the east. They were probably first introduced in Kent, with Lincolnshire and Norfolk following soon after.

Again the corn-mill is soon followed by industrial mills. The *Domes-*

Fig. 23.
Treadmill of the Middle Ages

day Book itself mentions a few stamping mills for crushing ore and hammer mills. Later we hear of paint-mills (1361), tanning mills (1217), saw mills (1376) and fulling mills (1168). By the thirteenth and fourteenth centuries every fairly-sized village had a mill of the new type, rented by a class of rentpaying millers. Conflicts between the lord and the tenant concerning mills begin to figure largely in legal documents. Also the mills in the river valleys played a large part in industrial revolutions such as the thirteenth century changeover from hand- or foot-fulling with water-driven hammers. This caused an emigration of the fulling trade to the country freeing the craftsmen from restricting local regulations (67).

We find a similar picture in Central Europe (68). By the eight century water-mills were well established in Thuringia, the Odenwald and Mühlhausen areas. The *Bohemian Chronicle* of Wenzel Hager claims that the first water-mill in Bohemia was built in 718 A.D. but this is now generally discredited. The emperor Henry I founded Goslar in 922 on a spot: "Where once a water-mill stood". The town of Augsburg built its "Lechwehr" about 1000, a weir to provide water to its mills. A water-mill with a mill-pond and water-race was built near the abbey of Viecht in the Inn valley in 1047 and an artificial water-conduit cut into the rocks on order of the emperor Henry III fed it. Herrad of Landsperg drew a corn-mill with undershot-wheel in her Hortus Deliciarum (1159). By the twelfth century they had penetrated into Scandinavia and they reached Iceland about 1200, about the period we hear of the first water-mills in Poland too. In 1243 mills were built near Breslau in the Oder and their rights were confirmed by the emperor Henry IV (1272). The first German paper-mill was built near Nürem-berg (69) by Italians in 1389.

The water-mill was applied to metallurgy early here, the new powerful hammer forges and water-driven bellows contributing materially to the increase in size and quality of wrought iron objects and to the new process which produced cast iron in large quantities for the first time. Thus the water-mill because one of the pillars of Central European mining and metallurgy. In this role it penetrated to regions still unconquered. By the twelfth century they were introduced in the copper-mines of the Harz and the silver-mines of Trient. They follow silver and copper mining into the Alpine regions, and are used by metallurgists and miners in Scandinavia (1530), Norway (1616) and Finland (sixteenth century). Water-driven forge-hammers are quite common in the thirteenth century, early in the fourteenth water-mills are used in the production of drawn metal wire. Water-wheels move mining hoists and grinding stones since the thirteenth century. Vienna mss. No. 5014 (1420) recommends the use of a water-mill to drill gun-barrels, so does another mss of 1430.

This wide-spread use of the water-wheel in industry contributed materially to the evolution of cog- and gear-wheels and to the practical aspect of mechanics as is shown in the engineers handbooks of the early sixteenth century. Unfortunately we know little about the con-

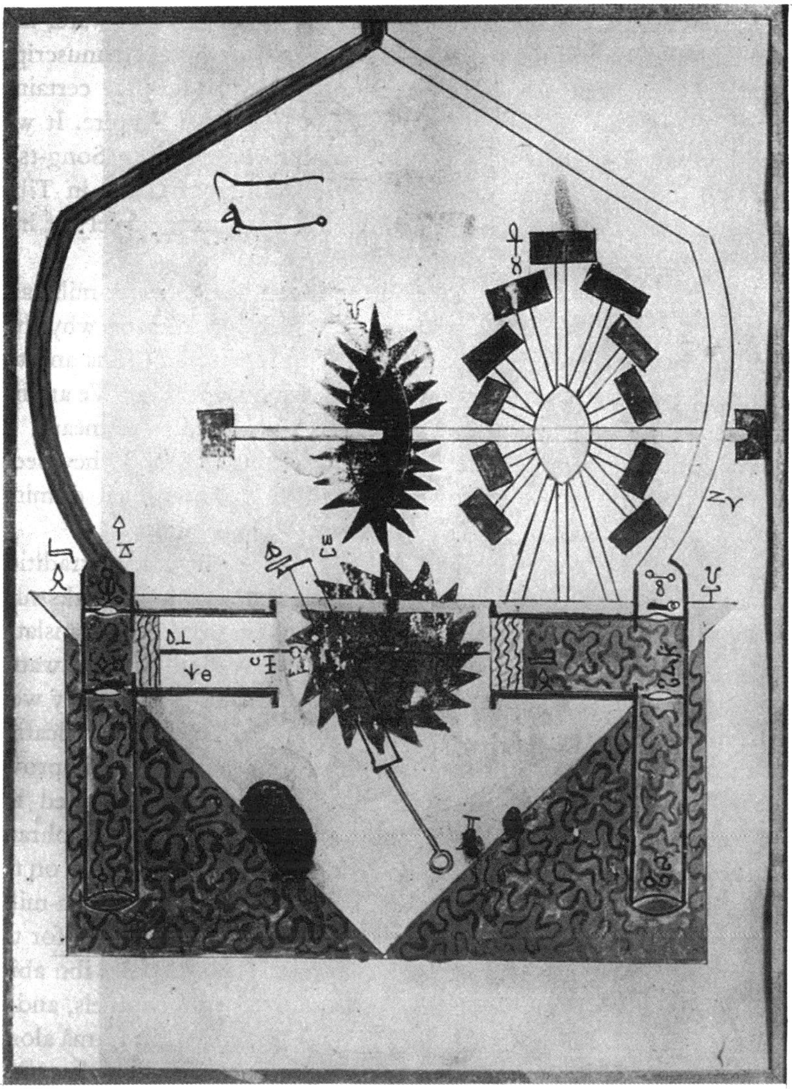

Fig. 24 Diagram of a water-mill in the treatise on Automata by AL-JAZARI

struction of medieval water-mills until the printed and illustrated handbooks show the enormous strides that had already been made since Antiquity. Cities and industries tended to move to running streams and rivers because of it, and thus make the transport of industrial products easier. It contributed materially to the rise of the cities and early capitalism.

We have only disconnected data on the history of the water-mill in the Near East and the Far East as many of the original manuscripts have not yet been translated or studied. The water-mill certainly travelled east during the later Roman and Byzantine Empire. It was well-known in China in the seventh century when King Song-tsen Gam-po (623—650) introduced the water-mill from China in Tibet together with butter, cheese and ceramics. The water-wheel for irrigation purposes came to Japan about 800 A.D.

Although Chinese annals ascribe the invention of water-mills and water-wheels to Tu Yu (222—284 A.D.) there is no reason why they may not have been derived from some region between China and the Roman Empire where they were well-known by that time. We are not informed whether the Norse or the Greek type of mill was meant.

The Talmud definitely refers to water-mills (70) and they seem to have been introduced into Palestine during the Roman domination.

In the world of the Islam there was of course a very strong tradition of Hellenistic and Roman engineering skill. Authoritative works such as those by PHILON, VITRUV, HERON and others were soon translated into Arabic. A Greek ambassador is said to have built a large water-mill at Baghdad (751). AL-JAGUBI (about 890) mentions that they were introduced in Baghdad a century before his time. The main application of the water-mill was irrigation. The Muslims added some improvements to the older constructions but they mainly sponsored the practical use all throughout their empire. Many cities on the Euphrates and the Tigris had floating mills and the coastal town of Basrah on the Persian Gulf even had its tide-mills, but they were mostly corn-mills. Syria was very famous for its water-wheels, more particularly for the big wheels in the Orontes at Antioch and at Hamâ. Qaisar ibn abî-l-Qâsim (died 1251) was especially famous for his water-wheels, and in the days of Abû-l-Fidâ's time (died 1331) there were 32 in Hamâ alone. It was formerly supposed that the crusaders impressed by these machines had introduced them into western Europe. However this revival of Hellenistic improvements merely coincides with the beginning of their fuller utilization in the west. The development is very different. When the water-mill becomes a major source of power in the west, AL-QAZWÎNÎ (1203—1283) can only describe different types of water-wheels in his *Geography* and AL-JAZARI mentions a water-wheel in his treatise of 1205 as one of the many amusing automata one could make. Apart from its use in irrigation the water-wheel never became the prime-mover in the Muslim world as economic conditions did not stimulate this development.

The windmills of the East

It is generally agreed (71) that the windmill was unknown to the Greeks and Romans. Hero of Alexandria describes a windorgan. The piston providing the air for the organ pipes is moved by a wheel "which has oar-like scoops like the so-called wind-wheels". The word used in the Greek text is "anemourion" (that is "wind-vane") which, except in this mss., occurs only in the writings of the twelfth century

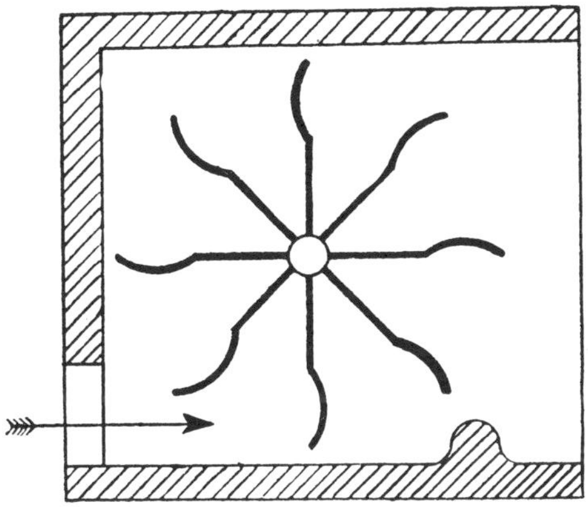

Fig. 25.
Plan of an eastern windmill

bishop Eustathius of Thessalonica. Some of the drawings in the Heron manuscripts show a kind of windmill moving the piston but they were later additions to copies of the original manuscript and must be considered interpretations by Christian (or Moslem?) scribes who knew the windmill proper. As no other classical engineer but Hero mentions anything like a windmill we must conclude that it was unknown but that the principle of the wind-motor seems to have been known though we do not hear of any application (72). Drachmann agreed that there is no evidence that "anemourion" can mean a windmill or wind-motor, or that Heron used it in this sense. He suggested that the "wind-wheel" was just a note in Herons' material for his (unfinished) book and therefore unknown to his contemporaries. But his suggestion may have inspired a much later inventor, who saw the manuscript and its illustrations to construct the Aegean windmill, the tower mill with eight sails.

The ancestor of the eastern windmill may be the well-known prayer-mill, in which scoops catching the impact of the wind rotate a vertical axle. Chinese travellers found it in use in Central Asia about 400 A.D. The first real windmill comes early in the history of Islam (73). During the reign of the Chaliph Omar I (634—644), thus we are informed by Ṭabarî, a Persian Abû Lulua (later his murderer) had claimed to be able "to build a mill that is rotated by the wind", and when he affirmed this on interrogation the caliph made him build one.

Our next information comes from al-Iṣṭakhrî, a Persian geographer (about 950) who revised the work of al-Balkhî "on the Figures of the Climates" and a similar passage, that seems to have been copied word by word, occurs in the works of Ibn Ḥawkal (74). They tell us of windmills in the Persian province of Seistan in the following words: "The region consists of evaporated salt-lakes and sand and it is hot. In it we find clusters of palm-trees and no snow falls there. The land is flat and one sees no mountain there. The nearest mountains are those in the direction of Farah. There strong winds prevail, so that, in view of these, mills were built rotated by the wind.

The masses of sand of this region wander from place to place. And if no precautions had been invented they would have flooded communities and cities. I have heard that, when they want to move sand from one place to the other without spilling it on the fields, which adjoin the sand, they enclose the sand by a structure like a fence of timber and thorns and the like higher than the sanddune. At the lower end they make a door. Through this the wind enters and blows the upper strata of the sand up like a whirlwind, so that the sand rises and falls far away where it does not harm."

We have quoted the latter part on the removal of sand to show that the inhabitants of Seistan seem to have been quite capable of directing air-currents at the early date. However, the main application of this skill was directed the building of windmills about which al-Mas'ûdi (947 A.D.) says (75): "Seistan is the land of wind and sand; the region is characteristic by the fact that there the wind turns mills which also pump water from wells to irrigate the gardens. There is no place on earth (and Allah alone knows) where people make more use of the wind." al-Qazwînî (76) reports about three centuries later: "There the wind never rests, so that in view of this they build their mills. They grind corn only with these mills. It is a hot country and it has mills using the wind." Then he goes on to describe how the inhabitants of this province remove sand, and his report is repeated by several other Arabic geographers (77). Hence apart from this local application of the wind to transport sand we find wind-mills in Seistan grinding corn and pumping water.

It would seem that this is the home-country of the eastern wind-mill and that its invention should be placed in early Moslem or even pre-

Fig. 26.
Dimashqi's drawing of a corn-mill

Islamic (Persian) days. The first description (78) of their construction (together with a drawing) is given by the Syrian cosmographer AL-DIMASHQI (1256-7—1326-7): "When building mills that rotate by the wind they (in Seistan) proceed as follows. They erect a high building like a minaret or they take the top of a high mountain, or hill, or a tower of a castle. On top they build one building on top of another. The upper structure contains the mill (raḥà) that turns and grinds; the lower one contains a wheel (daulâb = scooped waterwheel) rotated by the enclosed wind. When the lower wheel turns the upper millstones turn too. Whatever wind may blow, the mills rotate, though only one stone moves.

After they have completed the two structures, as shown in the drawing, they make four slits or embrasures (marmâ: loophole of a fortress) like those in walls, only they are reversed, for the wider part opens outward and the narrow slit is inside, a channel for the air, in such a way that the wind penetrates the interior with force like in case

Fig. 27.
Windmill in N. Seistan between Herat and the frontier, near Gourian
(Photo Ghirshman, 1952)

of the goldsmith's bellows. The wider end is at the entrance and the narrower end on the inside so that it is more suitable to the entry of the wind which penetrates the mill-house from whatever direction the wind may blow (hence the *four* openings in the structure). If the wind has entered this house through the entrance prepared for it, it finds in its way a reel (sarîs) like that on which the weavers wind one thread over another. This machine has twelve ribs (ḍil'), one could diminish them to six. On these fabric (châm = rough imbleached linen, etc.) has been nailed, like the covering of a lantern, only in this case the fabric is divided over the different ribs, so that each single one is covered. the fabric has a bump which the air fills and by which they are pushed forward. Then the air fills the next one and pushes it on, then it fills the third. The "reel" then turns; and its rotation moves the mill-stone and grinds the corn. Such mills are wanted on high castles and in regions, which have no water but a lively movement of the air."

The author then proceeds to tell us how sand is moved with the help of the wind. The passage quoted above gives us a very precise picture of an upper millstone rotated by a vertical shaft in its turn moved by a framework supporting (6-) 12 vertical sails. The drawing shows a hopper feeding the millstones through the central hole in the upper stone. The shaft is indicated to be of iron. A later historian tells us: "In Afghanistan all wind-mills and water-wheels are driven by the north wind and hence so orientated. Attached to the windmills there are series of shutters which are closed or opened to shut out or admit the wind. For if the wind is too strong the flour burns and even becomes black, even the millstone itself may grow hot and disintegrate." (79)

This type of structure seems very directly related to the Greek or Norse mill, which found its origin in the mountain range of N.W. Persia. The eastern mill would seem a Persian local adoption of the Greek mill to a region where there is no water but where steady winds prevail. Both the water-mill and the horizontal windmill are called "daulâb"! Subsequently the invention spread throughout the Moslem world and beyond it to the Far East in the twelfth century after having first been confined to Persia and Afghanistan. In China, India and the Moslem world they became important power resources used not only to grind corn and to pump water, but also to crush sugar cane, etc. As crushing mills they were a feature of the Egyptian sugar cane industry and thence travelled to the West Indies when Arabic experts from Egypt were lured there to help and establish the first sugar plantations. Due to the devastation of the irrigation works by the inroads and misrule of Mongols and Turks the windmill has now practically disappeared from Egypt and the Near East except for the country of its origin, Persia.

The Chinese still use for irrigation purposes and for the pumping of brine from deep wells a wind-mill combined with a chain pump which is derived from the Persian wind-mill. However, it is not built inside a mill-house but it stands in the open unprotected and it carries on its frame sails adapted from the Chinese fishing boats (80). They were first described by a Dutch traveller Nieuwhoff in 1655, but would seem to be much older. We are promised more details on these Chinese windmills in J. Needham's Science and Civilisation in China Vol. IV. 2.

The evolution of the western mill

There has been much speculation on the problem how the idea of the windmill came to the west. Some have contended that it passed from Persia to north-western Europe by way of the traderoutes through Russia and Scandinavia (71). However the sequence of the dates of the first windmills in the different European countries would sooner suggest that it might have come from the Moslem world by the way of Morocco and Spain. On the other hand the principle of the western mill is entirely different from its eastern counterpart, having a horizontal windshaft with four vertical sails from the earliest post-mill onwards. Hence it does not seem to derive from the Greek water-mill and it may therefore have a completely different origin. It has been derived by some from the Vitruvian water-mill and the choice proved to be a good one, as Smeaton later proved that it was much more efficient than the eastern mill.

However, its origin still remains a mystery. As the principle of the windmotor seems to have been known in the Hellenistic era it is just possible that this idea finally caught root and materialized in the form of the western windmill. The internal economy of the post mill is just like that of a water mill, but with the drive turned upside down; i.e. driving down from the sails instead of up from the water wheel. A millwright of the day familiar with the Vitruvian water mill could have devised a wind mill of the post mill type.

Again we would like to know more of the history of those quaint windmills of Crete and the Aegean islands (81), a few single specimen can be found as far west as Portugal. They have sails set like those of a boat and not spread on a lattice frame work. The shaft has 8 (-12) arms, slender round poles supported by stays of wire. Its sails are triangular with their base towards the circumference and their apex pointing to the centre. One edge is attached to the end of the pole, the other to a rope. While the mill is idle, the sails is twisted along the pole, when used they are unwrapped to a greater or less extent. The top of the mill can be turned towards the wind by using a long pole as a lever to move its conical cap supported on a ring of hard wood a little larger than that which caps the tower itself. Groups of these mills climb up the hills on the sky-line, though others are built on the flats near the sea. It has been suggested that the wind-motor and the sails of a ship are both at the bottom of the western mill. But these Cretan windmills are tower mills and we know of no tower mill antedating the post-mill. 13th and 14th century mss do not show jib sails which are found only in the Mediterranean area and the Iberian peninsula and always close to the sea. They seem later adaptions of the Western tower-mill to Mediterranean conditions. We would like to know more about the windmill built in Syria about 1190 by German crusaders (82).

The earliest mention in the West is found in a French charter of 1105 (83), which speaks of a "molindina ad ventum". It is still a novelty in twelfth century and yet possibly already more efficient than the older water-mill, for in the *Status de la république d'Arles* (1162—1180) the windmills pay a 5% tax and the water-mill a 3% tax on the turn-over. Here as in other European countries the wind-mill is primarily a corn-mill to become a prime mover at a later date. In Portugal the windmill is said to have been introduced by the Crusaders in the eleventh century which not only seems very early but less plausible in view of what we discussed above. In view of the different construction of the mills in East and West the Western windmill may well have been an independent European invention of the twelfth century.

The first definitely accepted reference to a wind-mill in England

Fig. 28.
The oldest and earliest wheat mill (wind), by PHILIPP GALLE

is found in the *Chronicle* of JOCELYN DE BRAKELOND (1437) describing the building of one by Dean Herbert at Bury St. Edmunds (1191) and its demolition by order of the bishop Sampson. A windmill at Weedley in the East Riding of York may date back to 1185 (84). The first English pictures in illuminated manuscripts begin to appear a hundred years later. They then begin to figure in works of art like the figure of a postmill is a Flemish memorial brass to Adam de Walsoken at Kings

Fig. 29.
The oldest and earliest wheat-mill (water), by PHILIPP GALLE

Lynn, he died in 1349. The massing of the present mills in the Eastern counties seems to prove that the windmill was introduced from the continent, possibly from the Low Countries though more probably from France where windmills are known to occur earlier. The Pope Coelestinus III (1191—1198) had to decide whether windmills had to pay the usual tithe so they were still a novelty at the end of the twelfth century. It has been assumed that some of the mills mentioned in the Domesday Book were windmills but we have no positive proof.

The windmill became the typical prime mover of the low plains reaching from eastern England via the Low Countries and the north German plain into Latvia and Russia. In its centre, in the Low Countries, the windmill is now primarily used as a prime mover of water-raising or pumping machinery (85). It has been claimed that the Cistercian monks introduced them during the large-scale drainage of the lakes and fens in these regions. The production of peat for the larger settlements here formed shallow lakes which grew larger in the period of 1000—1400. The systematic drainage of these lakes and fens started about 1300 after the encroachement of lakes and sea had been temporarily stopped by dykes and dams. Contrary to our expectation, however, the first wind-mills of the Low Countries were corn-mills. The oldest Dutch

Fig. 30.
The tjasker, a typical Dutch drainage mill invented by Cornelis Dircksz Muys in 1589

document is that of count Floris V who grants the burghers of Haarlem the privilege of paying but six shillings tax for a windmill and three for a horsemill, whereas non-citizens had to pay 20 shillings (1274). In the year 1299 the windmill of the former monastery Koningsveld at Delft was built and the books of the count of Gelre of the year 1294 show the building of a corn-mill at Logchem. Only about 1340 do we hear of the first drainage mill driven by the wind and not until about 1600 did it become a common feature of the Dutch drainage system.

In 1222 the town of Cologne in Germany built a mill on its town wall, a feature so common in late-medieval towns. Still the Dutch must have become clever mill-wrights soon for a mill at Cologne of 1392 was built by Dutchmen as was one at Speyer a year later. In 1237 we hear for the first time of an Italian windmill at Siena; in 1337 another one is mentioned (near Venice). Though Central Europe and Southern Scandinavia the windmill reached Finland about 1400.

The oldest type of windmill was the post-mill, in which the body containing the machinery is mounted on a high post supported by four diagonal quarterbars ending in two cross trees. It was a static mill orientated in relation to the prevailing wind. Its instability in high winds and the impossibility to use the wind from whatever direction it came led to the development of the tower-mill. It consists of a fixed

Fig. 31.
Water-wheel of 1288 in Sachsenwald (Germany)

tower which contains the machinery (may be even the house of the miller) while only the top or cap is turned to face the sails into the eye of the wind. It is usually supposed that it was invented about 1500 but this can not be true as we find on a tapestry at Nürnberg (1390) a tower of a castle from which tower protrudes a windshaft with four sails. Another tower mill is shown in the French mss *Livre des Merveilles* and in the Jewish *Haggadah of Nürnberg*, both of the year 1395. The latter mss. has a legend saying: "The wind in the mill grinds and not the force of the water." There is a 14th century French illuminated manuscript showing tower mills of a type recognisable similar to those found in certain areas of France today. However the connection of these early tower-mills and the later ones, and also their relation to the post-mill is not yet clear. Not until the fifteenth century does the windmill play any part as a prime-movers, all the examples mentioned are corn-mills. Its technical development seems to start with specialisation.

We end by mentioning a few suggestions for the use of the windmill which are less well-known. A Münich mss, of 1430 proposes to use a windshaft with sails acting as a brake when lowering heavy loads as blocks of natural stone. In the book dedicated by Walter de Millinate to king Edward II of England in 1326 the author illustrates the use of a post-mill to fling beehives into a besieged town. Lastly the Italian engineer Roberto Valturio (1413—1484) drew a design of a tank driven by two windshafts with sails.

Slowly the water-wheel and the windmill are improved by the craftsmen, but these prime-movers do not become significantly stronger until (after the futile attempts of Simon Stevin to calculate the power output of windmills) the production of better gear-wheels and the careful measurement of the efficiency of the various types of prime movers by Smeaton well in the eighteenth century, though by the impact of science they soon could be turned into the equivalent of 10—15 HP motors. However, their reign was already threatened for the new prime mover, the steam engine began to compete seriously around 1820 and displaced the older forms quickly after 1850. The old prime movers lost the battle against a new one perfected by cooperation of craftmanship and applied science.

Bibliography

1. Cranstone, B. A. L., *The Stone Age Man's Use of Power* (*Man*, vol. 51, 1951, 48—50); Sir Hartley, Harold, *Man's Use of Energy* (*Nature* vol. 166, 1950 368—376); Atkinson, R. J. C., *Stonehenge* (*Pelican Books*, A 450, London, 1960, 102—141)
2. Forbes, R. J., *Professions and Crafts in Ancient Egypt* (*Archives Int. d'Histoire des Sciences* no. 12, 1950, 599—618); Abd al-Mohsen Bakir, *Slavery in Pharaonic Egypt* (Cairo, 1952); Boak, A. E. R., *Manpower shortage and the Fall of the Roman Empire in the West* (Ann Arbor, 1955); Bark, W. C., *Origins of the Medieval world* (Stanford, 1958); Bilinski, B., *Le problème du travail dans Rome antique, l'époque royale et les premiers temps de la république* (*Archaeologia III*, 1949, 45—111; 415—425); Bilinski, B., *L'antiquité à la manière d'Hésiode. I. Le travail dans la Grèce antique.*; Westermann, W. T., *The slave systems of Greek and Roman Antiquity* (Philadelphia, 1955); Westrup, C. W., *Some notes on the Roman Slave in Early Times* (*Hist. Filol. Medd. Dan. Vid. Selsk.* 36, 1956, 3, 1—15); Vogt, J., *Sklaverei und Humanität im klassischen Griechentum* (*Abh. Soz. Wiss. klass., Akad. Wiss. Lit.*, Mains 1953, no. 4; *For man-power in ancient China see* Pulleybank, E. G., (*JEHSO* I. 1968, 185 ff); Hadjinicolaou-Marava, A., *Recherches sur la vie des esclaves dans le monde byzantin* (*Coll. Inst. Franc. Atchenès* XLV, 1950)
3. Mendelsohn, I., *Slavery in the Ancient Near East* (New York, 1949)
4. Westermann, W. L., *Industrial Slavery in Roman Italy* (*J. Econ. History* vol. II, 1942, 149—163); Bloch, M., *Comment et pourquoi finit l'esclavage antique* (*Annales, Economics, Sociétés, Civilisations* Vol. II, 1943, 30—44, 161—170); Yeo, C. A., *The Economics of Roman and American slavery* (*Finanzarchiv* (Tübingen) N.F. vol. XIII, 1952, pags. 445—485)
5. Lefebvre des Noëttes, *L'attelage, le cheval de selle à travers les âges* (Paris, 1931); Usher, A. P., *A History of Mechanical Invention* (Cambridge, 1954); Burford, A., *Heavy transport in classical Antiquity* (*Econ. His., Rev.* XIII, 1960, 1—18); see also pag. 136
6. Needham, J., *Central Asia and the history of science and technology* (*J. R. Central Asian Soc.* 1950, 139—141)
7. Kroon, H. M., *Hoefijzers uit de vroege Middeleeuwen* (*Tijdschr. Diergeneeskunde* vol. 53, 1926, 579—586); Michon, E., *Le trouvaille faite à Eauze* (*BSAF* 1927, 707—109); Winkelmann, Fr., *Über das Hufeisen* (*Germania* vol. 12, 1928, 135—143); Kaufmann, D. B., *Horeshoes in Antiquity* (*Class. J.* vol. 26, 1930/31, 619—620); Fraser, A. D., *Recent light on the Roman Horseshoe* (*Class. J.* vol. 29, 1933/34, 689—691); Méautis, G., *Les Romains conaissaient-ils le fer à cheval* (*Rev. étud. anc.*, vol. 36, 1934, 88); Carnat, G., *Le fer à cheval à travers l'Histoire* (Paris, 1951); see also chapter III, p. 156.

8. *Anth. Pal.*, IX, 418
9. Strabo, XII, 3.30, cap. 556
10. O'Reilly, J. P., *Some further notes on ancient horizontal watermills* (*Proc. R. Irish Acad.* XXIX, Sect. C., 1902/04, 55—84); Cecil, E., Curwen *The problem of early water mills* (*Antiquity* vol. 18, 1944, 130); Lucas, A. T., *The Horizontal Mill in Ireland* (*J. R. Soc. Ant.* Ireland LXXXIII, 1, 1953, 1—36)
11. Mayence, F., *La IIIe campagne de fouilles à Apamée* (*Bull. Musées R. Art. Hist.* Bruxelles, 1933, no. 1, p. 5 & fig. 5)
12. Vitruv, X.5
13. *de Rer. Nat.* V.516
14. *Nat. History* 18,23
15. Steensberg, A., *Farms and water-mills in Denmark* (Copenhagen, 1952, 294)
16. Bennett, R. and Elton, J., *History of Corn Milling* (London, 1899, vol. I); *On the early flourmills*, see also Forbes, R. J., *Studies in Ancient Technology* III, 138—146
17. Sagui, C., *La meunerie de Barbegal et les roues hydrauliques* (*Isis*, vol. 38, 225—231)
18. *Kosmographie*, edit. Wüstenfeld, vol. II, p. 381
19. Wiedemann, E., *Über ein arabisches eigentümliches Wasserrad* (*Mitt. Gesch. Medizin* vol. 15, 1916, 368—370)
20. Ausonius, *Mosella* V.362
21. Vitruv, X.1; X.5
    Strabo, XII. 3.30, cap. 556
    Pliny, *Nat. Hist.* 18.97
    Moritz, L. A., *Vitruvius' water-mill* (*Class. Rev. N.S.* VI, 1956, 193—196)
22. *Mostra Augusta della Romanità, Catalogi* (Roma, 1938, 3. edit., 591, no. 27); Jacono, P., *Ann. Lavori Pubbl.* 1939, no. 2, 217; Reindl, C., *Ein römisches Wasserrad* (*Wasserkraft und Wasserwirtschaft* vol. 34, 1939, 142, 143)
23. Benoit, F., *L'usine de meunerie hydraulique de Barbegal, Arles* (*Rev. Archéol.* (6) vol. XV, 1—40, 49—80)
24. Parsons, A. W., *A Roman Water-mill in the Athenian Agora* (*Hisperia* vol. V, 1936, p. 70—90)
25. Brett, G., *Byzantine Water-mill* (*Antiquity* vol. 13, 1939, 354—356)
26. *Mosella*, 361—364
27. Buckler and Robinson, *Sardis* vol. VII, 138, no. 169
28. Palladius, I.42
29. Ashby, Th., *The aqueducts of Ancient Rome* (Oxford, 1935. 46)
30. *C. Theod.* XIV.15.4
31. *Contra Symm.*, II. 949—950
32. *Bell. Goth.*, I.19
33. *Cod. Justin.*, XI.43.10
34. Cassiodorus, *Var.* III.31.2
35. *Digest* XXXIX 2.24
36. edit. Lindenbrogius, 1613, VIII.4
37. Salant, W., *Science and Society in Ancient Rome* (*Scient. Monthly* vol. 47, 1938, 525—535)

38. DRACHMANN, A. G., *Ktesibios, Philon and Heron* (Copenhagen, 1948); DRACHMANN, A. G., *The mechanical Technology of Greek and Roman Antiquity* (Copenhagen, 1963); REHM, A., *Zur Rolle der Technik in der Griechisch-Römischen Technik* (*Archiv. f. Kulturgesch.* vol. 38, 1938, 135—162); CHATLEY, H., *Engynes* (*Engineering* 1946, 388 ff); CHAPOT, V., *Sentiments des anciens sur le machinisme* (*Revue Etud. Anc.* vol. XL, 1938, 158—162)
39. SUETONIUS, *Vespasian* 18.2
40. CASSON, St., *Technique of early Greek sculpture* (Oxford, 1933); GUMMERUS, H., article on: "Industrie und Handel" (*Pauly-Wissowa* vol. IX, 1464—1490)
41. PLINY, *Nat. Hist.* 34.11; XENOPHON, *Cyropaedia* VII.2.4; AUGUSTINE, *Civ. Dei* VII.4
42. PAPPUS, Preface to Book VIII of the *Collection* (320 A.D.)
43. ZILSEL, E., *The genesis of the concept of scientific progress* (*J. Hist. Ideas* vol. VI, 1945, 325—349)
44. CORNFORD, F. M., *Was the Ionian philosophy scientific?* (*J. Hell. Stud.* vol. 62, 1942, 1—7)
45. TORREY, H. B., *The evolution of mechanical ideas in ancient Greek thought* (*Amer. Naturalist* vol. LXXII, 1938, 293—303); GOMPERTZ, H., *Problems and methods of early Greek Science* (*J. Hist. Ideas* vol. IV, 1943, 161—176)
46. PLATO, *Republic* 590c; *Laws* 807D; ARISTOTLE, *Politics* 1337 b, 9—14
47. DIJKSTERHUIS, E. J., *Maatschappelijke invloeden in de Griekse natuurwetenschp* (*Hermeneus* vol. XVIII, 1946, 23—29); CROISSANT, J., *Matière et changement dans la physique ionienne* (*Antiquité classique* vol. 13, 1944, 61—94); VERNANT, J. P., *Remarques sur les formes et les limites de la pensée technique chez les Grecs* (*Rev. Hist. Sci. Fr.* X, 1957, 3, 204—225)
48. O. KATTSOFF, LOUIS, *Ptolemy and the scientific method* (*Isis* vol. 38, 1947, 18—22); BÜHL, O., *Did the Greeks perform experiments* (*Amer. J. Physics* vol. 17, 1949, 384—388); GIGON, O., *Das Problem der Wissenschaft in der Antike* (*Universitas* vol. I, 1946, 1073—1084); LEVIN, S., *Socrates' rejection of science* (*Trans. Proc. Amer. Philol. Assoc.* vol. 79, 1948, 343—344); HORNE, R. A., *Plato and the rise of the anti-scientific sentiment* (*Physis* (*Ital.*) III, 1961, 4, 336—343); SANTILLANA, G. DE, *A proposito di reazione antiscientifica* (*Physis* (*Ital.*) IV, 1962, 2, 97—100)
49. BACON, FRANCIS, *Novum Organ.* I,1xxv
50. OERTEL, Fr., *Zur Frage der attischen Grossindustrie* (*Rhein. Museum* 1930, 230—252); FORBES, R. J., *The Ancients and the Machine* (*Arch. Int. d'Histoire des Sciences* no. 8, 1949, 919—933); WRIGHT, F. W., *Roman factories* (*Classical Weekly* vol. XI, 1917, 17—19); BIANCHI-BANDINELLI, R., *L'artista nell'antichità classica* (*Arch. Class.* IX, 1957, 1—17); FORBES, C. A., *The education and training of slaves in Antiquity* (*Trans. Amer. Phil. Assoc.* LXXXVI, 1955, 321—360); RANDALL, R. H., *Jr. The Erechtheum workmen* (*AJA* LXII, 1953, 199—210); WHITE, K. D., *Technology and Industry in the Roman Empire* (*Acta Class. S.A.* I, 1959, 78—89)

51. Rostovtzeff, M., *Social and economic history of the Hellenistic world* (Oxford, 1941, 3 vols)
52. Procopius, *Bell. Goth.* VIII.11,27; Vegetius, III.10; Zosimus, V.21.2
53. Thompson, E. A., *A Roman Reformer and Inventor* (Oxford, 1952)
54. Job 31.15
55. Bolkestein, H., *Wohltätigkeit und Armenpflege im vorchristlichen Altertum* (Utrecht, 1939)
56. Geoghegan, A. T., *The attitude towards labour in early Christianity and ancient culture* (Washington, 1945)
57. Thompson, E. A., *A Roman Reformer and Inventor* (Oxford, 1952)
58. Procopius, *Bell. Goth.* I.19
59. Venantius Fortunatus, *Carmina* III,12.37—38
60. Ausonius, *Mosella*, 361—364. The authenticity of this text has recently been discussed by Lynn White (Isis vol. 46, 1955, p. 291—292).
61. Bélidor, B. F. de, *Architecture hydraulique* (Paris, 1737, vol. I, lii, p. 321)
62. Seutonius, *Nero*, 31.2
63. Gille, B., *Esprit et civilisation au Moyen Age* (Paris, 1952); Id., *Le moulin à eau, une révolution technique médiévale* (*Techniques et Civilisations* vol. III, 1954, pags. 1—15); Sicard, G., *Les moulins de Toulouse au moyen âge* (Colin, Paris, 1953); Wilson, Paul N., *Watermills, an introduction* (London, 1955); Wilson, Paul N., *The origins of water power* (*water Power*, August 1952, 308—313); White, Lynn, *Technology and invention in the Middle Ages* (*Speculum* XV, 1940, 2,141—159); Bautier, A.-M., *Les plus anciennes mentions de moulins hydrauliques industriels et de moulins à vent* (*Bull. Philol. Histor. Fr.* II, 1960, 567—626)
64. *Vita Petrum* ch. XVIII
65. *Lex Salica* XXIX
66. Hodgen, M. T., *Domesday Water Mills* (*Antiquity* vol. 13, 1939, 261—277)
67. Carus-Wilson, E. M., *An Industrial Revolution of the Thirteenth Century* (*Econ. History Rev.* vol. XI, 1941, p. 39—60)
68. Reindl, C., *Die Entwicklung der Wasserkraftnutzung und der Wasserkraftmaschinen* (*Wasserkraft Jahrbuch* 1924, 1—27)
69. Johannsen, O., *Die erste Anwendung der Wasserkraft im Hüttenwesen* (*Stahl und Eisen* vol. 36, 1916, 1226—1228)
70. re̱ḥajim šel-lam-majim; Tos. Schabb. I.23,j. Schabb. 4a
71. Titley, A., *Notes on old Windmills* (*Trans. Newcomen Society* vol. 3, 1924. 41—51); Vowles, H. P., *An Inquiry into the Origins of the Wind-mill* (*Trans. Newcomen Society* vol. 11, 1930/31, 1—14)
72. Alexandrinus, Hero, *Pneumatica* I.34, edit. Schmidt, W., (Leipzig, 1899, vol. I, p. 205—207); Drachmann, A. G., *Heron's windmill* (*Centaurus* VII, 1961, 2,145—151)
73. Al-Tabari, Selections from the Annals (edit. de Goeje, Leiden, 1902, p. 1); Al-Mas'ûdî, *Prairies d'or* (edit. Barbier, Paris, vol. 4, p. 227); Muit & Weir, *The Caliphate*, Edinburg, 1915, 187
74. Al-Istakhrî, (edit. de Goeje, Leiden, 1927, p. 241); Ibn-Ḥawkal, (edit. de Goeje, Leiden, 1873, p. 299)
75. Al-Mas'ûdi, *Prairies d'or* (edit. Barbier, Paris, vol. 2, p. 80)

76. AL-QAZWÎNÎ, *Works* (edit. WÜSTENFELD, Göttingen, 1849, vol. II, p. 134)
77. ABÛ-L-FIDÂ', *Geographie* (edit. REINAUD, Paris, 1849, p. 340) (about 1300); YÂQÛT, *Geographisches Wörterbuch* (edit. WÜSTENFELD, Leipzig, vol. III, p. 42 (about 1200)
78. AL-DIMASHQÎ, *Manuel de la cosmographie* (edit. MEHREN, Copenhagen, 1874, p. 246)
79. GUZÛLI, *Matâli 'el-budûr* (Cairo, 1299/1896), vol. I, p. 50); MEZ, A., *Die Renaissance des Islâms* (Heidelberg, 1922, p. 439)
80. VOWLES, H. P. & VOWLES, M. W., *The Quest for Power* (London, 1931, p. 124); NOLTHENIUS, TUTEIN A., *Getij- en Watermolens in Vlaanderen* (*Tijdschrift v.h. Kon. Nederlandsch Aardrijkskundig Genootschap*, *Deel* LXXIII, No. 2, april 1956, pp. 162 t/m 166); NOLTHENIUS, TUTEIN A., *Getijmolens in Nederland* (*Tijdschr. v.h. Kon. Ned. Aardr. Gen.*, *Deel* LXXI, 1954, pp. 186 t/m 199); NOLTHENIUS, TUTEIN A., *Schipmolens* (*De Ingenieur*, Vol. 67, 1955, pp. A. 420 t/m 423); NOLTHENIUS, TUTEIN A., *Getijmolens* (*Suriname*) (*West-indische Gids* 1955, p. 219); NOLTHENIUS, TUTEIN A., *Getijmolens in Noord-Amerika* (*Orgaan Technische Ambtenaren v.d. Rijkswaterstaat*, Nov. 1955, pp. 113 t/m 116)
81. COBBETT, L., *Mediterranean Windmills* (*Antiquity* XIII, 1939, 458—461); BATHE, GREVILLE, *Horizontal Windmills* (Philadelphia, 1948); BAROJA, J. C., *Le moulin à vent en Espagne* (*Int. Congr. Europ. West. Ethnol.* Stockholm, 1951)
82. AMBROISE, *Estoire de la Guerre Sainte* (*Trans.* HUBERT, M. J., New York, 1941, 11.3227—3229)
83. *Magasin Pittoresque* vol. 20, 1852, p. 50; DESLISLE, L., *On the origin of windmills in Normandy and France* (*J. Brit. Archaeol. Assoc.* vol. 6, 1851, p. 403); GIRAUD, C., *Essai sur l'histoire du droit français au moyen âge* (Paris, 1846, vol. 2, p. 208); BIED-CHARRETON, R., *L'utilisation de l'énergie hydraulique* (*Rev. Hist. Sci.* 8, 1955, 53—72).
84. LEES, B. A., *Records of the Templars* (London, 1935, pp. 131, 135).
85. BOONENBURG, K., *Onze windmolens* (Amsterdam, 1949); STOKHUYZEN, F., *The Dutch Windmill* (Bussum, 1962); *The Principal Works of* STEVIN, SIMON, Vol. V, *Engineering* (edit. FORBES R. J.) (Amsterdam, 1965); see also page 62.

Chapter three

LAND TRANSPORT AND ROAD BUILDING

Introduction

Traffic on land was slower to develop than that on sea. This was partly due to the fact that the first urban centers arose in the Ancient Near East and the Indus valley, regions ordained by Nature for river transport. Early water transport also took away from the land those few goods traded over long distances and thus eliminated one of the most powerful incentives for an evolution of land transport. Prehistoric trade in Europe followed the big rivers as far as possible (fig. 32) (1). The Assyrian traders stationed in Asia Minor sent their goods down to the Euphrates or to the Tigris to be shipped home by water (2). The ancient Egyptian foreign trade with Crete, the Syrian coast and Somaliland or Araby was practically entirely coastal shipping to certain ports (3). Ancient Mesopotamian trade relied almost entirely on river transport (4).

Land traffic may hate and avoid the ascent and descent of hills and mountains, its real enemies are extreme cold and heat. Cold has made the extreme northern and southern regions of the earth sparsely populated, there roadbuilding becomes costly and uneconomic, generally speaking. The heat of the tropics and its deserts and dense jungles were formidable obstacles to ancient traffic until the advent of the railway, the motor-car and aviation.

Traffic and trade are related to a certain extent only. However, more powerful factors create long-distance communication and roadbuilding. A fanatical religion like Islam with its holy wars, its missionary urges and its pilgrimages promotes it. The rise of the early empires and their centralized organisations were based on quick information by letter and messenger. The mechanisation of armies which began with the introduction of the horse-drawn warchariot about 1500 B.C. demanded something better than a primitive track. A messenger service needed stations and rest-houses for men and beasts and supplies of water along the way at fairly regular intervals. The merchant followed the extension of power and his demands created a more stable and solid form of means of communication.

Upto that moment only human beings and messages made up traffic. Prehistoric trade is a barter from tribe to tribe. It never entails mass-produced goods but only such very valuable materials as precious metals and stones, pearls, amber, furs, silk, spices, salt and slaves. Later textiles and highly valued foodstuffs and other luxury goods were added. Even in classical antiquity such mass-produced goods like grain,

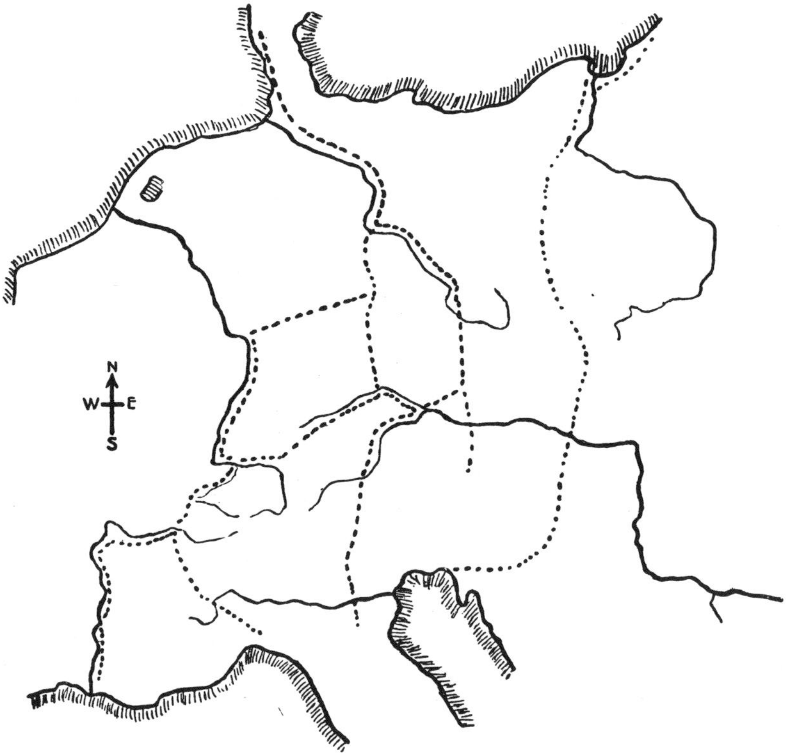

Fig. 32.
Traderoutes in Prehistoric Europe

olive-oil or wine took the way of the river and the sea, for every 100 miles of land transport doubled their price.

From the point of view of traffic the Mediterranean world was greatly privileged. Climate, situation and indentation are ideal for the evolution of long-distance communication (6). To the north of the bordering mountain-ranges there are the north-European, Russian and Asian plains with their gold, slaves, furs and corn, which appealed to ancient trade. To the south it was enclosed by the deserts of the Sahara, Nubia and Arabia, which are broken by the Nile valley (and the rather unnavigable Red Sea), the gate to India and the East for the ancients. Though divided in three distinct zones by the peninsulas of Italy and Greece, it has four important gateways to regions beyond, Iberia, the Black Sea, Egypt and Syria. Mesopotamia and Persia with its second set of communications to India and the East belong to the Mediterranean world rather than to Asia.

Its mountain-ranges (generally east-west or northeast to southwest) have steep slopes, deep valleys lead into highlands poor in rain. The short rivers with strongly varying volume of water are hardly navigable and of little use to water-wheels generally. This is particularly true of North Africa and Asia Minor. Agriculture in the eastern part is dominated by irrigation, in the west "dry farming" dominates. The highlands were quickly denuded not only through human agency but mainly by the herds of goats held by the ancients.

In Antiquity the centre of this world, which was only united by the Roman Empire, moved from the east (Alexandria) to the west (Rome) and back to the east again (Baghdad) with the advent of Islam. But wherever its focus was, nature had shaped it for sea-traffic. Even the Romans, who were unwillingly drawn into naval warfare with Carthago in which they invented the corvus to suit their army tactics, dominated their empire by being masters of the Mediterranean. Hence land-traffic and its conspicuous result, road-building could only gradually gain ground by a series of strongly centralised political powers, the Persian Empire, the Hellenistic monarchies and the Roman Empire. In Roman times Western Europe had hardly more than 5—12 inhabitants per km^2 against 18—27 per km^2 in the Mediterranean world. Therefore even in Roman Europe the network of roads, dictated primarily by strategic considerations, was never as dense as in the south and the east, where over 50 million Roman subjects lived.

Land-traffic in the Persian Empire

Long before the Persian king of kings started to develop a consistent traffic-policy tracks and bridle-paths had connected the settlements of men all over the world. In certain cases they had grown into systems like the ridge-ways and trackways of prehistoric Britain (7) which were levelled and traced to a certain extent and which converged on Salisbury Plain with its religions monuments such as Stonehenge, Woodhenge and Avebury. Hence they may not only have served a modest trade on pack-animals but also may have been pilgrim-ways. In other parts of prehistoric Europe tracks had been constructed through swamps and marshes. These log-roads avoided great detours (8). In prehistoric Malta the strange artificial ruts cut into the rocky soil seem to have served to guide the carts loaded with earth and pulled inland by man-power to the terraces on the hillsides (9). Recent practical tests have proved that these ruts were not made for wheeled vehicles but for slidecars, consisting of two wooden shafts supported at the front end by the draft animal (horse or ox), while the rear-end trails along the ground. The shafts widen towards the "heels" and the body of the "car" usually made of wickerwork hangs somehwere in the middle of the shafts. The ruts do not only lead from the valleys to the hills but some keep to the hills, others to the low grounds and hence they also seem to have been used for the transport of goods between settlements. The Egyptians had built causeways and roads from the quarries to the Nile for the transport of building materials. In Crete a tracksystem with guardhouses across the island from Gortyna to Knossos may date back to 2500 B.C. (10).

In the urban centers of the Near East attempts at road-building had been made for centuries and most of the main streets and market squares

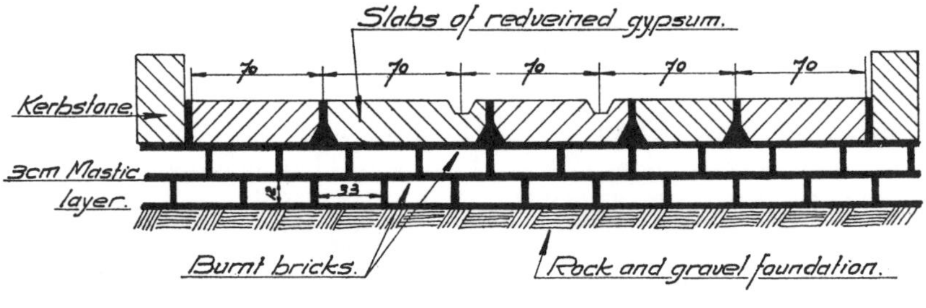

Fig. 33.
Procession-road in the temple of Ishtar at Assur

of Egyptian and Mesopotamian towns were paved with flagstones and slabs often brought down from the mountains at great cost. Some of these pavements are properly constructed on a brick foundation, the slabs being set in bitumen mortar, specially designed joints preventing the bitumen to sweat out and spoil the pavement in summer-time (11). Some of these paved streets are "processional roads" connecting temples in the city with the "festival-houses" (temples) outside the city-walls. On these roads carts carrying the statues of the gods were drawn between

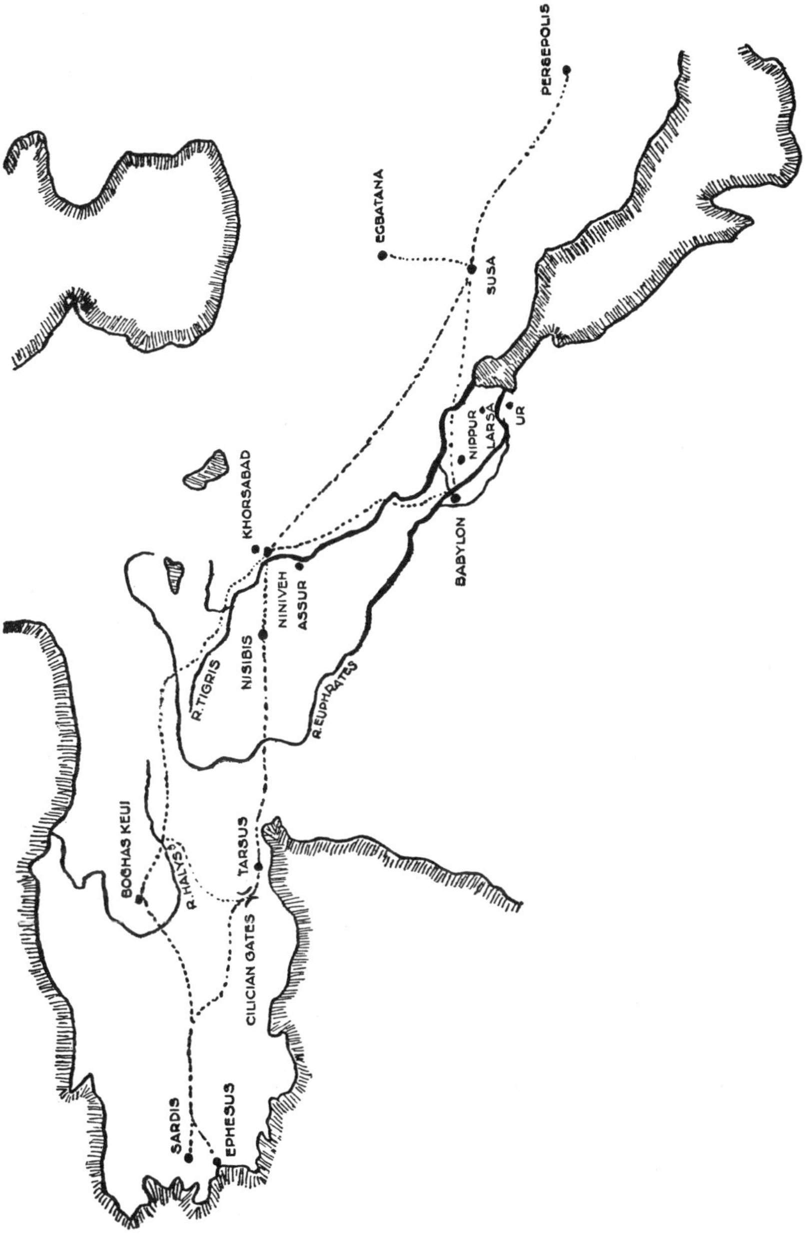

Fig. 34.
The main Persian highways

these temples on the occassion of religions festivals. The oldest processional road was found at the capital of the Hittites, Hattusas, in Asia Minor. It dates back to 1200 B.C. Next come a series of such processional roads ("tallakti") in the Assyrian Empire at Assur (700 B.C.), Babylon (700 & 600 B.C.) and Uruk (300 B.C.). Some of these pavements have artificial cartruts, the gauge of which coincides with those of prehistoric Malta and Greece (gauge 138—144 cm). It seems that the cart drawn by two oxen is responsible for this "standardisation".

However, this meant little for long-distance communication. Even the great coastal road from Egypt to Gaza, Syria and Mesopotamia was little more than a track, on which wheeled traffic was virtually impossible because of the stepped ascents and descents (ladder-ways, Gr. klimakes) in the defiles of the Nahr-el-Kelb and other places. The Assyrian attempt at world domination meant a certain degree of mechanisation of the army, which from now on had a siege-train. Already Tiglath Pileser I (c. 1100 B.C.) had an engineers corps, the "ummani", who laid pontoon bridges and levelled tracks for the carts and siege-engines (13). He tells us about one of his campaigns in the mountains of Elam: "I took my chariots and my warriors and through their wearisome paths I hewed a way with pickaxes of bronze and I made passable a road for the passage of my chariots and my troops." (14) More often, however, his army met "a difficult terrain, where my chariots could not pass", and the army would have to proceed on foot.

Neither did Sargon II, Sennacherib (c. 700 B.C.) and their successors build a network of highways. Though no concentrated effort was made there were certain attempts to establish regular communications with distant parts of the empire. The Assyrians more or less regularised the system of communicating by fire signals, an old system already in use in Israel (15) and Mesopotamia (16). Now, during the hours of the night "stacks of wood were fired and torches which shone at a distance of a double hour (10.269 m)" (17). There was also some form of messenger service for the kings, "men of the letters of the king" served him in the empire, and centralised messenger-services were at his disposal in the main towns. Maps of towns and town-quarters and lists of towns and countries with their distances in "double-hours" were at the disposal of these couriers. Some of these lists take the form of the later Roman itineraries such as the list of towns from Assur to the Persian Gulf and their distances (c. 800 B.C.) (18). We know that the tracks in the Assyrian empire had roadsigns "which allowed travellers to find their way after dark". On desert tracks there were "guard-stations at regular distances and wells were sunk" (19). Ferries (nâbiru) made the crossing of large rivers possible in certain cases bridges (tilûru, gisru) were built such as the one at Babylon, over which classical authors waxed enthousiastic (20).

Though Sanherib remarks that "roads should be well kept and streets well constructed" and Essarhaddon in rebuilding states that it would be his policy "to open their roads throughout the land so that they can resume traffic with all neighbouring countries" this meant little more than levelling tracks and policing them. It did not improve on the speed of travelling since the news of the rise of the Nile at Elephantine reached to Memphis with a speed of about 11 km per hour (c. 2000 B.C.) or Hammurabi could write to an official at Larsa (200 km away): "Day and night you shall travel, so that you may arrive at Babylon within two days." (21)

Though the Assyrians knew different types of four wheeled carts for the transport of building materials, a kind of covered waggon (for women?) and two-wheeled chariots none of these played an important part in land-traffic. Moreover the comparative study of ancient chariots and waggons is still in its infancy (21a). However, in this period falls an important change, the use of the horse as a riding animal. After the horse was introduced in the Ancient Near East (c. 2000 B.C.) from the North it had only served to draw war-chariots. Both in the Near East and the Far East (22) men had ridden on donkeys, but the donkey as a riding animal plays no part in the story of the road. Since the days of Sargon II, however horses were occasionally ridden, and his army had cavalry (710 B.C.), the earliest traces go back to the days of Assurnasir-pal (c. 875 B.C.) In Egypt it was still a matter of surprise that a prince of Saïs rode on his charger and "asked not for his chariot" (c. 750 B.C.). Sometimes the later Assyrian kings send horsemen (ragab) as messengers to foreign countries (23). This art of riding horses came from the North and the East (Persia) and for the first few centuries it was still confined to nobility. Only much later did the horse affect the development of road-building, as few merchants used it.

The horse came into its own after Cyrus (550—530) founded the Persian empire, for the Persians were a horse-loving and horse-riding people. The Persian kings tried to get a firm grip on a subject by amalgamating them into one nation. A good stable currency for the whole empire was one of their means to achieve this and so was the messenger-service on which the classical authors inform us (24). This "arrangement was devised to meet the huge size of the empire and enable the king to learn with great celerity the state of affairs at any distance. Cyrus first ascertained how far a horse could travel in one day without being overridden, and then he had a series of posting stations (stathmoi) built, one day's ride apart, with relays of horses and grooms (hodophylakes) to take care of them, and a proper man is in charge of each station to receive the dispatches and hand them on, take over the jaded horses and men and furnish fresh ones. Sometimes we are told this post did not even halt at night, the night messenger relieves the day messenger and rides on."

We are informed that the stations (hippones, stathmoi) were some 25 km apart (25). They also housed troops and seem to have played some part in a system of fire-signals (26), which the Persians also used in their wars against the Greeks (27). The perfection of this messenger-service (angareïon) is laid to Darius I (521—485), echoes of it can be found in the Old Testament mentioning "posts on horseback, and riders on mules, camels and young dromedaries" (28). It was a Royal postal-service and its command was held a great honour. In fact Artaxerxes entrusted Codomannus (the later Darius III) with it (29).

This track-system led to the palaces of the Persian kings at Susa, Persepolis and Ecbatana. Its main track, which Herodotus calls the Royal Road, leads from Sardis (the most western provincial capital) and the harbour Ephesus to Susa via Issus, Laodicea, the Cilician Gates, Tarsus, Zeugma, Nisibin and Niniveh. (30) This stretch of 2600 km was covered by messengers in nine days, ten times as quick as an army. The distance between Babylon and Susa was covered in $1\frac{1}{2}$ days, that is an average of 150 km a day. It should be realised that this is a speed attained by specially equiped messengers but it certainly was not beaten until well after Napoleon's days.

This trackway-system was no attempt at large-scale roadbuilding. We do hear of two paved highways between Susa and Ecbatana (31). But Diodorus tells us at the same time that stretches of the Royal Road were difficult because of the many steep slopes but still practicable for carriages and Xenophon in his story of the retreat of the 10.000 Greeks has only too many stories of beaten paths and chariots stuck in the mud (32). In fact the tracks were simply cleared and kept in repair by the ancient simple means described in a report of an address of Cyrus the Younger to his army (33): "You gentlemen, who are in command of the roadmakers, you have the lists of soldiers I have disqualified from serving as javelin-men, bowmen or slingers, and you will make the old javelin-men march with the axes for falling timber, the bowmen with mattocks and the slingers with shovels. They will advance by squads in front of the waggons so that if there is any roadmaking to be done, you may get at work at once, and in case of need, I may know where to get the men I want." Roadbuilding here simply meant levelling a track, and that was its meaning in the East until well in the Roman times.

Apart from perfecting postal services the Persians were also responsible for the general introduction of the dromedary and the camel as a riding animal and a beast of burden (34). The camel was domesticated in Central Asia in late neolithic times, at the same time the dromedary was tamed and bred in Arabia. The Egyptians by their contacts with the desert-dwellers knew the dromedary from pre-dynastic times onwards and so did the Jews from the time of the patriarchs. About 1100 B.C. the camel was introduced from Media by the Assyrians for mountain traffic; they adopted the dromedary for desert travel about 700 B.C. The basis of the ancient southern Arabian kingdoms was camel-breeding. The true camel, an animal of temperate zones, though introduced there about 1100 B.C., was never a succes in Arabia. The dromedary, however, became the most important means of desert transport until the advent of the motor-car in our era.

The rise of camel-transport as a more international form of traffic began in the Syrian desert (35). Its rise became possible when the two horns of the Fertile Crescent were held by one political power or at least two friendly interdependent ones willing to take the short cut through the desert for trade's sake. Also the Arabs should by then be sufficiently advanced in camel-breeding to have sufficient riding- and pack-animals available to trade such specific agricultural or coastal products as salt, on which trade Palmyra was to grow rich. It was important that routes be traced with a perennial water-supply from springs or wells, no toll-bariers, the chronic threat of desert robbers firm in hand. As a matter of fact the early Arab tribes had their intertribal wars and supplied warrior-bands to their allies in the fertile Crescent but left the traders of Syria and Mesopotamia alone in the earlier centuries of desert trade. Desert robbers are first mentioned in the third century A.D. (36).

The above conditions were first fulfilled in Persian times. Tadmor may appear in inscriptions of Tiglath Pileser I, but the first traceable desert tracks date of the Persian period. The keys to this trans-desert traffic between Egypt and Syria on the one hand and Mesopotamia on the other hand are Palmyra and Petra, but their rise is due to different political causes. Palmyra flourished in the Persian period but it suffered a decline in the Hellenistic period when the direct route from Syria to southern Mesopotamia was deserted until well in the second century B.C. Then Palmyra grew in importance again, especially when the Romans and Parthians came to a mutual and advantageous understanding about the importance of desert trade on the eve of the first century A.D. These desert roads were retraced from the air and this is not the only field where air archaeology has proved invaluable for retracing the ancient network of roads and tracks (37). The Romans developed a series of highways throughout Palestine, Syria and Transjordania well-marked with milestones as "feeders" of these desert routes, which remained unpaved but well-marked and supplied with wells or cisterns every 30 miles (38).

Palmyrene trade reached its apex when Petra declined and Armenia became a battlefield between Romans and Parthians. Its decline came with its fall in 271 A.D. but there was a short revival after the reign of Justinian VI until the early Muslim period. The revival in the eighth century lasted some 500 years but dropped off sharply after the sack of Baghdad (1258) and ended with the fall of Constantinople (1453).

Petra first became important in the Persian period and its prosperity rose with the rivalry of the Ptolemies and the Seleucids (second century B.C.) which tended to diverge all trade with S. Arabia to Damascus and Syria instead to Egypt where the Ptolemies tried to draw it. Pompey's conquest of Syria awakened Petra's interest in the Palmyra road which interest was maintained until the reign of Trajan. Early in the Christian era Bosra began to take over Petra's role as the "desert port" of Arabia Felix and Petra declined. The Persians may therefore well claim the honour of having initiated a new form of traffic, dromedary transport through the desert, which ever since was important in the connections of the Mediterranean world and Europe with the Far East and with Africa where the camel was introduced in Roman times.

Traffic in the Greek World

The coast of the Greek homeland was strongly indentated and nearly every Greek town could smell the tang of the sea. "The Hellenes engaged together in no enterprise before the Trojan war on account of weakness and lack of intercourse. And they were united only when they were making considerable use of the sea." (39) Even "those who dwell more in the interior and away from any trade route" (40) sought the shortest route to the sea. Even in STRABO's days the roads of Greece were generally bad and hardly ever drained (41). This frustrated development of roads was partly due to the fact that mountain-ranges upto 3000—5000′ cut the country up into sharply divided valleys. More important still was the lack of political unity which persisted upto Roman times. When ARISTOTLE wants to describe the political structure of Greece he has to review the laws of some 150 city states or poleis. This lack of a strong central power may well be the reason for the bad road-system. When unity was finally enforced by foreign domination, Greece was no longer an important factor in the economy of the ancient world and none of the great Roman highways touched the Greek homeland.

We have little direct information on Greek road-building (42) but we are sure that it was not due to Phoenician influence, formerly much overestimated. The Mycenean civilisation of Greece, heir to the older Minoan traditions had indeed constructed some "cyclopean" roads in the region of Argolis and Mycenae. These roads have a pavement of polygonal slabs, well-drained foundations and several bridges (43). These may represent the "well-drained roads" mentioned in by HOMER (44) for we hardly find any well-built road in Greece before Roman times. This is clear from many passages in PAUSANIAS' description of Greece (150 A.D.) where many roads are proclaimed to "grow steeper and more difficult to man on foot," "a mere footpath", "easier for men than for mules" and "impassible for carriages by reason of its narrowness" (45). Only in a few cases they are "suitable for carriages". On the very important route-from Megara to Corinth he says "it was first, they say, made possible for footpassengers by (the mythical) Sciron, when he was war-minister of Megara; but the emperor Hadrian made it so wide and convenient, that even chariots could meet on it."

In fact most Greek roads were hardly more than footpaths or bridlepaths, mostly hardly suitable for pack-animals, as steep slopes were taken by steps cut into the rocks (klimakes, basmides). Carriages existed but were unpopular. Their use was considered effiminate and even forbidden in certain cases, such as processions from Athens to Eleusis. A kind of covered waggon (kanadron) was used almost exclusively by women and priestesses. Carts (hamaxa) were used in certain localities for the transport of grain, marble, and the like if the roads permitted it. In a few cases special roads were built for such purposes from the mines and quarries to the coast (45a). There was one to the quarries of the famous Pentelic marble in Attica, well-paved with slabs, and with holes along the side which probably held poles to prevent the sledges or carts with quarried marble from sliding from this regularly sloping road. However, such roads built for special purposes can not be quoted as proofs that the ancient Greek roads formed a good means of promoting land-transport and even the much better Roman road-system never invited a really important road traffic or long-distance haulage of goods important in everyday life such as cereals and the like. There was a limited land-transport for luxury goods, but generally speaking the rule held, that larger quantities of heavy goods were transported as quickly as possibly to the nearest navigable point of a river or to the coast. Another one with artificial wheelruts led from the white marble quarries of Agrilesa to Kamaresa.

Road signs were formed by piles of stones. Hesychius, commenting on these roadmarks tells us that Mercury after killing Argus had to throw an "absolution stone" on the roadside. This custom of throwing a stone on a pile at certain points of the road after saying a prayer still exists in many countries. In historical times square stone pillars with one or more busts of Hermes, god of the travellers, and called "hermae" were erected at such spots. The Greeks seldom resorted to the planting of trees along their roads. PAUSANIAS mentions only the roads along the Isthmus which ran between "rows of pinetrees most of them shooting straight up into the air "and that along the Alpheus near Herea" planted with myrtles and other trees" (46). However there was a general sprinkling of resthouses and inns along these tracks for "roads without inns are no better than life without holidays' (47).

Road building was therefore seldom considered a serious task of the state, though in Sparta the two kings "were to judge in all matters concerning highways" (48) and Peisistratos of Athens made a traffic law for Attica and had the roads measured and repaired (550 B.C.). These special surveyors, called bematistai, mapped the roads from a central point, e.g. the agora (market place) of Athens (49). Actual road building was even in disrepute for "Epaminondas (of Thebes) undertook the survey and inspection of all roads which the citizen who disliked him had entrusted to his care to insult him for this work was in descredit. However by his excellent administration he made many Thebans compete for this job as a high honour" (50).

However, there was one exception, the sacred road. This pavement with artificial wheelruts already attracted the attention of early travellers in Greece. "The term rut must not be understood in the sense of a hole or inequality worn by long use and neglect in a level road, but of a groove or channel purposely scooped out at distances adapted to the ordinary span of a carriage, for the purpose of steadying and directing the course of the wheels and lightening the weight of the draught on rocky or precipituous ground in the same way as the socket of our railroads. Some of these tracts or stone railways, for such they may in fact be called, are in a good state of preservation, chiefly when excavated in strata of solid rock." (51)

These sacred roads connect a city with a sanctuary and they usually represent the journey of a god on earth. Such roads as that from Athens to Eleusis, Sparta to Amyklai or Elis to Olympia were used for pilgrimages and religious festivals. Others lead to the famous oracle of Delphi.

The ruts were carefully hewn, polished and levelled so as to form a perfectly smooth and easy track for the cart wheels. The levelling instrument used was probably the dioptra. The Greeks considered the rut (ogmos) the real road and sometimes left the rest of the road surface unpaved. When these wheelruts were used by ordinary traffic difficulties might be expected if there was only one pair of wheelruts. One was lucky if one could say with Ion: "I seldom was molested, from my track no villain drove me." (52) In case of two chariots meeting each other, one had to give way, leaving the ruts to stand aside. This might become a reason for a violent quarrel like the one in which Oedipus killed his father, unknown to him, on the Cleft Way to Delphi (53). Some of these roads have points (ektropai) built in at certain distances.

The gauge of these ruts (usually 7—10 cm deep and 20—22 cm wide) was 138—144 cm like those of Maltese and Etruscan roads, though no connection has been traced between these systems. A few ordinary roads had such ruts too, sometimes even double tracks like that from Athens to Delphi, Sparta and Elis and Athens to Pyrgos. In the latter case a low wall crosses these tracks, it may have been a toll-barrier where the goods were transferred. Some of the sacred roads were entrusted to religious authorities such as the Amphictyones, the board of the Delphi oracle whose law of 380-379 B.C. says: "...and the bridges the Amphictyones shall repair, each his own part, and they shall see to it that they are not damaged, and the ambassadors of the Amphictyones shall look after the roads, whatever is needed for them, and they shall punish (those who damage their roads)." (54)

Ancient Greece never had anything like the Persian postal service. The Greeks used fire signals occasionally (55) and most cities had two or more fast-sailing ships (56) for quick long-distance communication. Some cities also had special messengers or runners (hemerothromoi, dromokerykes) which in time of war served as spies. Merchants had slaves which they used as runners.

A factor in this story of Greek roadbuilding was the gradual deforestation of the homeland. In the days of the Persian wars "the stay of Xerxes in Pieria lasted for several days during which a third part of the army was employed in cutting down the woods on the Macedonian mountain range to afford his forces a free passage into Perrhaebia." (57) When Macedonian power grew and its armies had an engineering corps with Thracian road-engineers (58) and a siege-train, it became necessary "to build fortresses and cut straight roads for military purposes" (59). The Greek armies now had their war-engines and became convinced of the necessity of good roads when they marched into Asia with Alexander. With the Persian tracksystem Alexander's army could make an average of 30—33 km a day but on certain occasions even marched 74 km. This meant that even when the central power of the king of kings declined the satrap or local governors kept the roads in good repair.

Even when Alexander's empire was cut up into four major Hellenistic monarchies this lesson was well remembered and the Persian postal service maintained notably in Egypt. The Zenon and Hibeh papyri (60) confirm this statement of classical authors (61). It was definitely a state-service (62) headed by officials called cleruchs. The land-mail service included a series of stations headed by a horographos where horses are cared for and messengers could rest. Apart from this there was a river-mail service. The land service ran two routes east and west of the Nile and sometimes took private letters. Apart from this there was a service of camelriders (camelites) and one of runners. The service was financed by taxes levied. This "liturgy" sometimes took the form of exaction of animals and fodder from the local population or requisitioning of houses. Later the right of the official to requisition was reduced and certain classes of priests exempted from this tax.

There never was any appreciable road-travel in Hellenistic or Roman Egypt, though a time-table of the camel-post was found (63). The ease with which passengers and goods could move up and down the river made roads superfluous. The camel, now universally used for landtransport was antagonistic to paved roads. Nor were any large bodies of troops stationed in Egypt, and a military road-system was never developed. However a road-system was developed and improved in Hellenistic times, that of the Nile to the Red Sea which was so important for Alexandria's trade with Arabia, India and the Far East. Traffic sailed up the Nile to Coptos and then struck eastward to the coast of the Red Sea of old (64). The oldest road led straight to Koseir (Albus Portus) where the ships could then be loaded, thus avoiding the long and dangerous passage through the gulf of Suez. Later more northern ports like Myos Hormos and Kenah (Kainopolis) were used, though the Roman traffic often struck south along the coast at Koseir to reach the port of Berenice. These roads were in reality all tracks but fortified watering stations (hydreumata) were provided along the route. These were hostels with substantial rubble-walls and flanking towers at the angles and at either side of the gate way. In the centre was an open space for the animals and a well. Most had brick reservoirs lined with cement for storing water. The Coptos-Koseir road was further marked out by a system of intervisible beacons or signal towers.

In the other Hellenistic monarchies the ancient systems were kept more or less in repair and changed little until Roman times. The old Persian roads of Asia Minor (65) still ended in the few available western harbours. The harbour of Miletus got silted up first, Ephesus followed in the Middle Ages. In the meantime, however, traffic became centered on Byzantium and the main routes now cut the province diagonally. Even under Turkish rule the postal service was still maintained.

Evolution of the Roman roads

What failed to be achieved by the Greeks was realized by the Romans who inherited from them the complex of ideas belonging to the maintenance of city life. The city (polis) came to be characteristic for the form of group-selfexpression of the Western Mediterranean and it was supplemented by the legists and civil engineers of Rome. With the stubbornness and tenacity of the peasant, which they always remained at heart, the Romans achieved an excellent land-transport system (66).

When Bergier wrote the first modern handbook on road-building in 1622 (67) he believed that the Phoenicians and Carthaginians had instructed the Romans in this art. This opinion based on passages from later encyclopedists (68) have been repeated frequently (69). Isidorus tells us: "A street (strata) is so called because it is worn by the feet of the public." Lucretius says: "And the streets of the roads already worn smooth by the feet of the public." The same "strata" also means "laid down", that is spread with stones. Now the Carthaginians are said to have been the first to pave the roads with stones; afterwards the Romans laid them out over almost the whole, for "directness of journeys and to prevent the populace being workless". But not being a roadengineer he confuses embankment, pavement and other elements of roads and then introduces startingly modern theories about road-policy, which have proved all wrong.

It is true that the Carthaginians paved their streets and made a few tracks and embankements (70) but the true impulse came from Italian soil. The Romans learnt their townplanning from their Iron Age ancestors of the Po valley, whose settlements contain log roads and other primitive forms of pavements (71). The Etruscans taught them the art of engineering, drainage and road-building (72). Though some have discussed and even depicted very sophisticated types of "Etruscan" roads (73), it now seems quite clear that most of them are just well-drained tracks on which the later Roman roads were built (such as the V. Aurelia, V. Clodia and V. Cassia) (74). Some of these roads must have been well passable for carriages, like the one on which the Vestal Virgins travelled from Rome to Caere when the Gauls invaded Italy (75). Some of these roads had artificial wheelruts and shunts like the Etruscan limestone paving slabs discovered at Fiesole, Perugia and Graviscae (76). They had also invented the slabs with a series of grooves, called "cordonato" by the modern Italian, which anti-slip pavement they lay in steep streets (e.g. at Fiesole). Finally the Romans used the extensive knowledge of building materials of the Greeks and adapted their lime-mortar and their "macadam" city-pavement (lithostratos). These three elements the Roman engineers welded into an excellent art of their own.

The Romans therefore inherited a system of tracks such as the later Via Gabina, Via Latina and those already mentioned (77). The most important of these was the Via Salaria which carried the prehistoric salt trade from the sea coast to the Sabine highlands (78). It was first paved as far as Reate, but Augustus paved it as far as Truentum on the Adriatic. Also there were cattledrifts like those from the meadows of Samnium to the Apulian lowlands. Most of these tracks disappeared in the course of history, being used as a foundation for later roads. Thus the Via Appia partly covered the old gravel road (V. Norbana) from Velitrae-Cora to Norba.

The need for better roads arose from the expansion of Rome about 500 B.C. Conquest and colonisation want hand in hand. The new colonies of Roman citizens, often rebuilt older towns, and the reconciliation policy extending the rights of citizenship (partly or wholly) to the conquered, who were of Latin stock, formed strong bonds which held in the Gallic invasion and the Samnite wars. The founding of Velitrae (494 B.C.) and Antium (467 B.C.) started a movement lasting until the days of Ceasar. The roads were gradually gravelled, then paved.

In this earlier phase these roads were not primarily strategic. They served traffic and communication, they followed the conquest but only

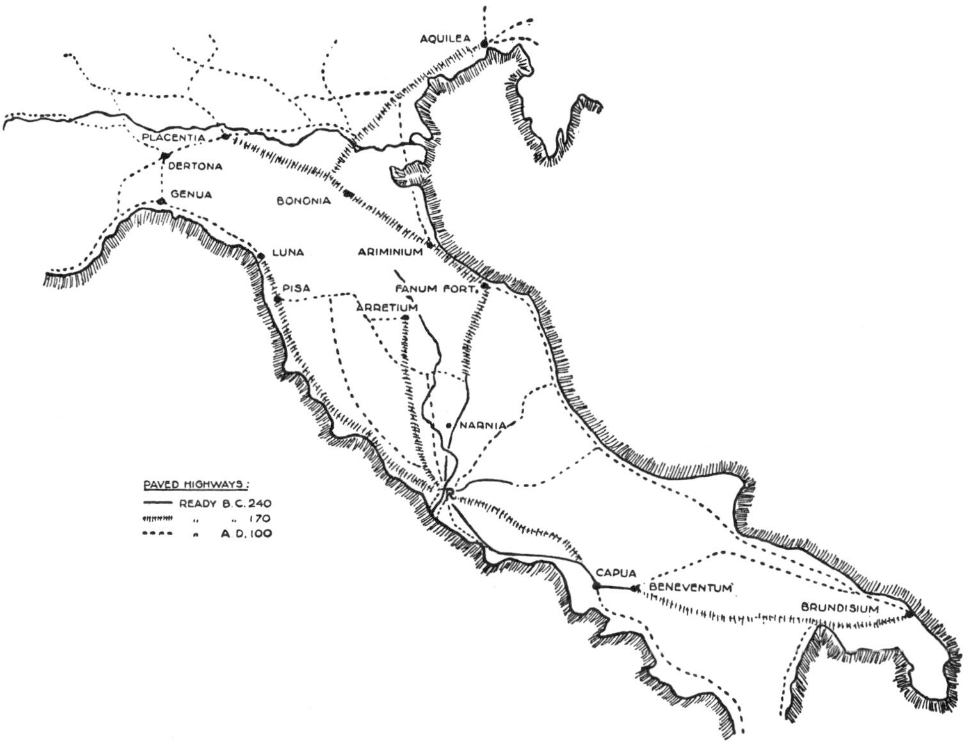

Fig. 35.
Growth of the road-system of Roman Italy

many years after the region was pacified and safe. Thus ten years elapse between the fundation of Placentia and Cremona and the building of the Via Aemilia. Only in later phases do we find such definitely military roads as those along the "Limes" of Germania and Syria. Other factors besides colonisation hastened rebuilding of the tracks. The growing cultivated area and the growth of landed property now shut in the old undefined track.

Technical improvements like the introduction of lime mortar from the Greek cities in southern Italy (300 B.C.) and the natural cement called "puzzolano" (150 B.C.) hastened this process. The flagstone pavement could now be made watertight and rested on a solid foundation. The first of the great highways, the Via Appia, from Rome to Capua was paved by the blind censor Appius Claudius (312 B.C.) in order to bring the Campania and its harvests within reach of the old city. It began as a gravelled road but after the Samnite wars the section to Bovilae (296—293 B.C.) was paved with the "fines levied on cattle breeders" (79). In 289 B.C. the oldest stretch was widened and two sidewalks were constructed. Shortly after the invasion of Pyrrhus the Via Appia was gradually extended to Brundisium and Tarentum.

Shortly after the Gallic invasion Gaius Flaminius divided up the "ager gallicus" and hence the Via Flaminia had to be extended beyond Spoleto to Fanum Fortunae (220 B.C.). The twenty years at the end of this century, because of the Second Punic War, show a lull in the progress. Then the defeat of Hannibal inaugurated a period of virgorous expansion. The valley was conquered and the colonisation of Bologna and Aquilea meant the building of the Via Aemilia. Around 175 B.C. the V. Aurelia and V. Cassia followed, financed by the Carthaginian war indemnity. The Via Postuma followed and the west coast road from Aquilea to Brundisium. At the same time the streets of Rome were paved and "the roads outside Rome were covered with gravel and footpaths raised at the sides." (80).

Then the three Macedonian Wars gave the Balkan Peninsula to Rome. In 148—145 B.C. on the farther coast of the Adriatic opposite Brundisium an extension of the V. Appia, the V. Egnetia, was built reaching from Dyrrhachium to Thessalonica and later beyond to the Hebrus (81). It was later connected with the new Greek highways constructed under the Empire. Nero built the road from Byzantium to Thrace, Trajan and Hadrian paved roads along the westcoast of the Black Sea to Tomi (82).

The Civil Wars inaugurated another period of slackness. Still G. Gracchus succeeded in getting his *Lex Sempronia viaria* adopted (123 B.C.). It ordered the extension of the Italian roadsystem, with more attention to be devoted to drainage and foundation problems, and also increased the number of milestones. In the meantime the occupation of most of Asia Minor led to a repair of the existing Hellenistic system by M. Aequilius (129 B.C.). In the Gracchian period puzzolano, a natural volcanic trass found near Puteoli, was discovered. It promoted the application of concrete constructions on a much larger scale. The consul Marius extended the road along Italy's west coast (V. Aemilia Scaurus) from Luna to Genoa, which together with new roads in the Po valley now connected Italy through southern Gaul with the new roads in Spain from Gades (Cadix) through Numantia to the Pyrenees. The link from the mouths of the Rhône (Arles) to Genoa ran over the Mont Genevre (Alpes Cottiae), only during Augustus' reign was the present Riviera road (V. Julia Augusta) was built (13 B.C.).

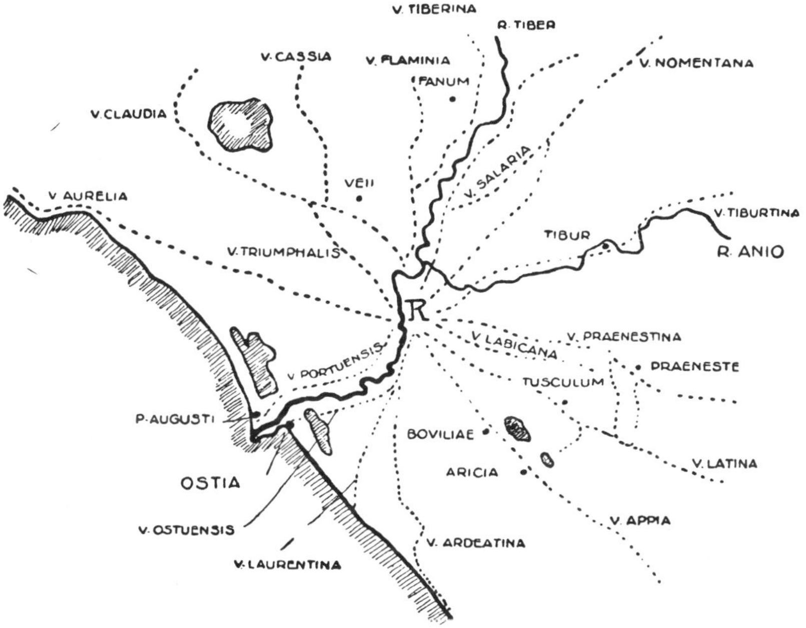

Fig. 36.
Roads radiating from Rome

Caesar's Gallic Wars started organised roadbuilding in Gaul. Before the Roman Conquest there were many native tracks but the great highways belong to the Empire. Their natural centre was Lugdunum (Lyons), whence a road south along the Rhône led to Arles and the coastal roads. A western road running through the country of the Averni reached the mouth of the Loire and then struck south to the present Bordeaux. A road to the north-west reached the coast opposite Britain, the north-eastern road led to the valley of the Rhine and Germania. These large roads formed the backbone of an intricate web of roads built by Claudius, Trajan, Nerva and Vespasian. Trajan connected this system with the roads leading through Germany to the Danube and the Black Sea.

Already in Republican days two roads penetrated the newly-won territory of Spain. However Augustus laid out a proper road-system comparable to Agrippa's plans for Gaul. He surrounded Spain with a great circle of roads with one or two branches penetrating to the inland mining districts of Rio Tinto, Almaden and the Sierra Morena. One of the chief functions of these roads was conveying the ores and metals to the seaports for shipping to Rome. The frequency of relaying roads points to intensive traffic and the determination to exploit the resources of the district. Older tracks were rebuilt into important trade-routes which were kept in good repair as is evident from van SICKLE's (83) data compiled in Table I.

Claudius' occupation of Britain resulted in the building of durable commercial highways, partly trackways rebuilt. London remained the centre whence trade went to the Channel ports. Military roads were built in Wales and the Lake District as well as along the Wall. No milestones earlier than Hadrian survived, during whose reign the roads along the north Welsh coast were taken in hand (83a).

Africa, one of the granaries of Rome, was completely pacified by Augustus, who founded colonies, constructed roads and provided the province with police forces. Then whole triangle Carthage-Cirta-Tacape needed good communications between the agricultural districts and the seaports and roads for the patrols of the military police (84). More to the west there were towns belonging to Spain rather than to Africa for the traffic went mainly by sea as there seems to have been a break in the road-system along the coast.

In Syria and Palestine the old roads were gradually paved (84a) and provided with hill forts along the desert against brigandage. This gradual construction of well-built roads started by Herod and his family was crowned when Trajan annexed Arabia Petrae (Transjordan) and built a splendid system of roads with a line of watchtowers and posts. The roads of Syria and Judea were carefully paved and garrisoned, the road from Antioch to the East constantly repaired and in northern Syria miles of roads of black basalt were built which have survived almost intact. The trade from Arabia to the north was sheltered by the Strata Diocletiana during the rise of the Sassanian Empire which military road was well-equipped with forts, water-supply from wells (85) and camps for the Bedouin auxilliaries.

In Asia Minor a consistent road-system was only possible after the Romans give up the system of buffer-states between themselves and the Parthians. Thus its roads were mainly built by the Flavians, Vespasian, Titus, Nerva and Hadrian. Here again we find many military roads on the protected border, the "Limes", which system the Flavians also adopted on the Rhine and Danube frontier. The road-system on the Rhine was begun between 73 and 84 A.D. but the "limes germanicus" was not completed until the reign of Trajan. The log-roads built by Caecina and Domitus from the Rhine eastwards into the territory of the German tribes were also military roads to combate the danger of raiding tribesmen. Thus, at the zenith of its power, the Roman Empire disposed of a 90.000 km system of paved highways, 14.000 km of which were located in Italy proper. If we include the secondary, tertiary and gravelled roads we obtain a total of 300.000 km. Twenty-three highways radiated from the capital, Rome, and led to the confines of the Empire. The topography of these roads has been studied in great detail (86) and new finds constantly increase the total mentioned above.

Roman materials and road-constructions

Unfortunately we possess no Hellenistic or Roman handbooks on road building. Even Vitruvius (87) is rather vague on the paving of streets though he is more explicit on floors which were composed of four strata, as confirmed by excavations. Vitruv's instructions are: "If we must lay our floor (ruderatio) on level ground, we must inquire whether the soil is solid throughout; it is then to be levelled and upon it the foundation (statumen) carrying another layer (rudus) shall be laid. But if it is a made site, in whole or in part, it must be very carefully rammed with piles." After discussing the building of upper floors and the planks to be used he continues: "Next a layer of stones (statumen) will be laid each of which it at least fist-size. When the statumen has been laid the rudus, a mixture of hardcore and lime, shall cover it. If fresh bricks are broken up for use, one should use three parts on one part of lime; if it is of old materials, five parts of rubble are to be mixed with two of lime. Let it then be laid on, and rammed down with repeated blows by gangs (decuriae) of men using wooden stamps. When the stamping is finished it must be not less than nine inches thick.

Upon this, a nucleus of powdered pottery is to be laid, three parts to one of lime, forming a layer of six inches. On the finishing coat, a pavimentum is to be laid to rule and level."

Excavations have proved that roads too consist of four different layers (statumen, rudus, nucleus and pavimentum or summum dorsum). The statumen can be compared with our handlaid foundation, the rudus is a watertight layer shutting out the damp from the statumen. Thus statumen + rudus together form the foundation proper. The nucleus is perhaps intended to serve as a "binder" for the pavement. The Roman authors (89) and poets have little say about roadbuilding except Statius, the poet-laureate of the emperor Domitian, who describes the building of the Via Domitiana (90 A.D.) in these words (90):

"Now the first stage of the work was to dig ditches and to run a trench in the soil between them. Then this empty ditch was filled up with the foundation courses and a watertight layer or binder and a foundation was prepared to carry the pavement. For the surface should not vibrate, otherwise the base is unreliable or the bed in which the stones are rammed is too loose. Finally the pavement should be fastened by pointed blocks and held at regular distances by wedges. Many hands work outside the road itself. Here trees are cut down and the slopes of hills are bared; there the pickaxe levels the rock or creates a log from a tree; there clamps are driven into the rocks and walls are woven from slaked lime and grey tufa. Hand-driven pumps drain the pools formed by underground water and brooks are turned from their courses."

However, the Roman road-engineer never stuck rigidly to this four-layer scheme. He shows great skill in varying it to local conditions, in using local materials and constructions to achieve his goal and he was never dogmatic to his own disconfiture. The type of construction was adapted to the type of traffic. It will be clear that a "transport road" like the one constructed by Hadrian for the transport of Numidian marble from the quarries at Simitthus to the harbour of Thabrace will differ from the simple country track. Such tracks or "viae vicinales", just levelled, rolled and drained, existed along many superior types of roads. We know little about the Roman gravel roads because most of them have disappeared by now. However in the days of Palladio (1518—1580), famous architect of the Italian Renaissance, methods were still in use reputed to be Roman practice (91). Probably all secondary roads (viae vicinales) were at least gravelled. The course of the road was levelled, sometimes rammed or even rolled with wooden or stone rollers (92), then an embankment of sand with a gravelled surface or a gravel layer was laid. Stones from the neighbouring fields served as curbstones. On both sides there were ditches down to 6′—8′ below the road surface.

Paving the road surface meant limiting the width. Thus the 120′—100′ old track gradually became the 40′ metalled road surface as decreed by Augustus for great highways (viae munitae). The delimited road, the "agger publicus" was state property. The shoulders, often 6—9 m wide, enclose the "actus" or carriage road and its sidewalks (iter) for foot passengers, which were often separated by a low stone wall (umbo) or a strip of grass. Gravel roads of this type survived in France under the name of "chaussées de Brunhaut" or "chaussées en accôtements".

The Romans also used cobble roads, laying the cobbles in a 10 cm sand-bed. Most cobble-surfaces survived in city pavements. A primitive macadam road-type was imported from Greece. Like this "lithostratos" the Roman macadam road was made by spreading chips, taken from

Fig. 37.
Normal Roman Highway (1 :85)

20—40 cm	stone setts or gravel concrete
30—32 cm	concrete with crushed stone and trass
30—50 cm	slabs and blocks in cement mortar
20—30 cm	mortar layer on top of sand course

the stone quarries, on the subsoil and ramming or rolling (ingenti cylindro) them to compaction. Repairs consisted of a fresh layer of chips rammed into the original suface.

As soon as lime and puzzolano were known mortars for grouting and concrete mixes could be used. Pavement slabs were usually cut from lava (silex) or tufa, hence the term "silicarii" for paviours. A green variant (cappellaccio) of tufa, was very popular and so was peperino, a conglomerate of volcanic ash and gravel. Bricks (testae) or tiles (tegulae) were also used, pounded tiles forming the main ingredient of the concrete (caementum) used. Even forge cinders or slags were used to surface the so-called "viae ferrae" many examples of which were found in Britain (92).

Lime mortars were used to grout the joints of paving slabs and also gravel layers. The varying qualities of different sands were early realised. VITRUVIUS proposes the following sand-lime ratios for river sand, down sand and quarry sand: 2-1; 2-1 and 3-1. If gravel (glarea), chips or crushed bricks were added to the mortar (usually three to one) the common concrete (caementum) was obtained. For road-building a gravel concrete was very popular with a mortar consisting of one part of lime, three parts of puzzolano and $2^3/_4$ to $3^1/_2$ parts of sand. This strong hydraulic mortar compares very favourably with modern portland cement mortars. As during the Empire other deposits of trass, besides Puteoli, were found and exploited, e.g. in the Eiffel district.

The usual type of main highway consisted of a 20—30 cm statumen of hand-laid slabs, a 30—50 cm rudus of grouted cobbles or crushed stones, a 30—50 cm nucleus of rich gravel concrete and a 20—30 cm pavimentum of 30—100 cm square stone slabs or a rich trass-concrete with aggregate upto 6 cm. This total depth of 100—140 cm is some 3—4 times that of a modern roadcarpet and hence the Roman roads have aptly been called "walls on the flat" (93). Even with the relatively small traffic (though dense locally) they stood up to the wear of iron-tyred wheels and hoofs of unshod animals for 70—100 years before they needed thorough reconstruction as shown by the data from Spain, though on more frequented stretches a 30—40 years' service is more probably. We have few figures on the extent of traffic and the frequency and costs of repairs. Though the Romans used perfectly sound building methods their invention of concrete halted the development of more plastic road carpets like cobble and macadam roads. Their rollers were not heavy enough for compaction of such roads. The care for drainage of the foundation was excellent and thus saved the rigid thick concrete mass from breaking up by frost or water. The pavement had no expansion joint and thus was little suited to climates of extreme temperature changes. The use of local materials meant dissimilar materials in different stretches of the road subjected to the same hammering action of hoofs and tyres.

In the northern parts of the Empire cracks and potholes formed more easily and thus in the later Empire constant repair took the overhand over construction of new roads. The cobble and macadam roads probably had a life of 10—15 years. There were of course many variants of the common constructions mentioned here (95) but the above figures give averages obtained by the comparison of ancient and modern roads (96). The width of a 40′ carriage road plus side-walks andshoulders is about 80′, the other roads have 12′—16′ paved surface on a total width of 36′—40′. On some military roads the side-walks were used by carts, the main road being used by the marching legions. The ditches (sulci) were usually 10′ wide and as deep as the foundation course, though often they go 10′ deep. The usual camber is 1:20. Milestones, the invention of which was ascribed to G. Gracchus (97), were of course much older. The Roman milestones, the oldest of which go back to about 250 A.D. have made the study of the Roman highway system much easier (98).

Disposing of efficient surveyors' instruments the Romans could set out (99) their roads perfectly straight as far as natural conditions would permit. CODRINGTON found that in Britain they seldom deviated more than half a mile on 20—30 mile stretches (100). In general inclinations are gelow 6%, though some of 8% occur.

Peculiar constructions were found all over the Empire. Artificial wheelruts (101) were made in certain Alpine passes, the Balkan and the Alpine provinces and in certain parts of Gaul, the gauge being again the same as in early Malta and Greece. Cuttings and embankements were very common. The Via Appia ran through a cutting driven 120′ into Mount S. Angelo near Terracina (102) and near Arricia 43 m retaining walls support the same road. Part of the road through the Medway valley near Rochester was built on a foundation of oak piles (103). In the Kasan pass and the Grében defile on the Danube the road was constructed on a wooden scaffolding let into the rock. Both Tiberius and Trajan made such roads, a picture of which occurs on Trajan's column. The structure has now disappeared and the road is recognisable only as a ledge in the rocky shore (104). On part of the Via Tiburtina, in order to protect it from the waters of hot sulphur-springs, the road ran in a concrete channel, the outside of which was protected by travertine slabs. The log-roads, first mentioned as "pontibus palude" by Caesar (105) were built on the right bank of the Rhine by Domitius and Caecina (106) after earlier examples built by the prehistoric tribes north of the Rhine (107). These few instances will show the ingenuity of the Roman road-engineer to cope with difficult local conditions.

CONTROL AND FINANCING OF ROAD AND TRAFFIC

Many of these roads were built by the army. The excellently organised Roman legions and their army train (impedementa) with large quantities of artillery like ballistae and catapultae needed good roads. Hence in peace-time the army and its engineering corps (the "centuria accessorum velatorum" formed at the end of the Republic) built roads as well as bridges, camps and forts. This was particularly true in Africa and the

Danube and Rhine provinces. During Augustus' reign there were several revolts because of this enforced road-building. Claudius also used gladiators for roadbuilding, slaves and captives were put to this kind of work too.

About 200 B.C. the provincial authorities began to procure additional labour and they gradually took over the greater part of road-building, charging contractors (redemptores) with the actual work and retaining only the supervision. Such contracts were very popular during the Empire but also the source of many frauds. In A.D. 21 "the state of the public roads was made the subject of a debate by the senator Gn. Domitius Corbulo. The highways, he said, were in a bad condition throughout Italy, neglect everywhere and in some places impassible. He attributed the mischief to the fraudulent practices of the contractors and the inattention of the magistrates. He came to superintend the business, but the advantage, whatever it was, that accrued to the public, did not counterbalance the ruin of individuals, who suffered both in reputation and fortune, by the harsh decisions of Corbulo and in the confiscation of their effects." (108) Claudius later punished Corbulo (109). In later centuries the authorities often ressorted to forced labour or exactions of money for road building (110).

During early Roman history the task of building roads was entrusted to individual censores, who like Appius Claudius or Flaminius often attached their name to this particular road. Later on emperors, generals or rich private individuals often did the same, prompted by rewards such as special medals, statues, memorial stones or even the right to erect a triumphal arch. Then special aediles called curatores viarum were installed as temporary functionaries, special senators often being entrusted with the supervision of traffic. As legislation on public and private roads became more complicated (111) the emperor Augustus created special permanent boards (22 B.C.) called curationes viae directly controlled by the emperor for special roads, districts or towns. Nero created the lower rank of subcurator. Many local authorities served on these boards particularly during the later Empire when road-building became less important than repairs and the army left this to the authorities. The provincial roads were administrated by the provincial governor and his nominees, but during the later Empire all these functions were taken by state officers.

Financing these roads was an old problem. The earliest roads were often financed from the booty taken from foreign conquests. An agrarian law of 111 B.C. wanted to make all those living along the road contribute towards its building and upkeep. In 50 B.C. the senator L. Scribonius Curo failed to push a tax on wheeled vehicles. Usually building and repairs were paid by the treasury from direct taxes. In certain cases emperors or private individuals face the costs. Thus Augustus pays most of the Via Flaminia but leaves part to the triumphant general C. Calvisius Sabinus who then builds his triumphal arch at Rimini. Private individuals or governors have left us word of such acts on milestones (112). Certain towns like Nuceria were granted tolls to raise sufficient funds for road-building (113). The same forms of financing applied to secondary roads (114).

Data on the actual costs of roads are scarce. The curator A. Vibius Temundinus repaired a 20 mile stretch of the V. Caecilia, regravelling part and re-paving the rest of the road in the Apennines at the cost of 150.000 sesterces (£4500.—) about 80 B.C. (115). The emperor Hadrian repaved part (15 miles, 750 passus) of the Via Appia "paying himself 1.147.000 sesterces, while the owners of the neighbouring lands had collected 569.100 sesterces" which works out to something like £3.000 a mile (116). The same emperor entirely rebuilt this road from Beneventum to Aeclinum (20 miles) at the cost of £18.000 per mile including the price of the land. In view of the size of these "walls" these prices seem comparatively low.

The roads served the army and state officials for whom the state post (cursus publicus) catered. They were also the servants of commerce but one should not exaggerate the volume of Roman trade and commerce in which much of the traffic was carried on only part of the year. Only the navigable rivers and the sea provided cheap avenues to the markets. The Mediterranean was the finest asset in Rome's imperial economy, even if the Romans always disliked the sea and spoke little of sea-transport (117). The motives of travel in Antiquity differed from ours (118). Apart from the displacements of army troops there was a large amount of official travel and business trips. Sightseeing was not common in the earlier periods. The ancients did not appreciate nature as we do and abhorred the mountains. They went to seaside resorts for their health and later in the Imperial period only tourism to famous towns and monuments became popular, mainly for instruction and education. This is particularly noticeable in Egypt in the second century A.D. However there was a tax on road travel there varying according to the rank of the traveller (119), the tariffs for courtesans were practically prohibitive. This money went into road-building funds.

The provincial governor, in the case of Egypt the prefect (120), had to provide the traveller with a passport. The local authorities surveyed and ran local hostels for officials, but there were many other forms of inns and hostels (121). In the case of important officials travelling abroad special instructions were given and requisitions were ordered (122). Scientists and students travelled to different universities such as that of Alexandria famous for its doctors of medicine (123). Then there were of course pilgrims going to different shrines. Some of these travellers succeeded in getting permission to use the official postal service but most of them had to provide for their own means of transport.

The "cursus publicus", the state postal- and messenger service, was the backbone of this passenger transport. By it the emperors could rule from Rome by letter alone according to classical authors (124.) Its origin goes back to the third century B.C. when officials tried to keep in touch with each other and with Rome. Requisitions for travelling facilities were even then possible (125) for senators. Later Caesar gives written permits to travellers and transports. The provincial governors had their messengers (statores) and the taxfarming companies their own couriers (tabellarii) (126). However, we find no definite traces of the postal service said "to have been instituted by Gaius Graechus" (127) though there was some postal service between Rome and Rhegium (128).

The emperor Augustus completely organised and centralised such earlier services (129). The Ptolemaic system seems to have inspired him for he tried to finance it by making the cities and population along the roads pay for it (130). It was first confined to the neighbourhood of Rome but gradually gew to embrace the whole empire under Trajan. Gradually the old system of financing became impossible and Hadrian made it a state-organisation. Private tabellarii were not allowed to use its facilities, which served officials only who had to get an official passport (diploma). However soon private travellers were allowed its facilities as a favour, animals and ships were allotted to private individuals and at the hostels presents lay awaiting them. At regular intervals laws had to be enacted to combat such abuses.

This combined postal-, passenger- and goods-service had strict regulations. Along the main highways there were stops where horses could be changed (mutationes) and larger stations and hostels (mansiones). The mansiones were about one day's journey (37 km) apart. On rivers there were ferries, and the sea routes were covered "dromones" or "cursioriae", fast sailing ships. The diploma rigidly prescribed the type and amount of transport one was allowed as well as the hospitality in the mansiones. For travel on the secondary roads or branch road one needed a special "evectio".

The heavy clabulariae (angariae) drawn by oxen carried a maximum of 1500 Roman pounds each (492 kgr) of army goods, parchement and papyri, products from the imperial estates, etc. The number of oxen or horses for this slow cursus clabularius was 8 in summer and 10 in winter, lower officials were allowed only this form of transport, also soldiers if travelling far and all sick people. Express goods and precious metals were transported by the cursus velox which used several types of carriages: 1) the rheda (vehiculum, carpentum), a four-wheeled cart with mules (8 in summer, 10 in winter) and a maximum load of 1000 lbs (330 kgr); 2) the carrus, a four-wheeled cart carrying 600 lbs (198 kgr) max.; 3) the verreda, originally used by women only, which was drawn by 4 mules and could transport 2—3 persons and max. 300 lbs (99 kgr.); and finally 4) the two-wheeled birota with 3 mules carrying 200 lbs (66 kgr) and 1—2 passengers. The cursus velox also had riding horses (veredi) and pack-horses (parhippi) which carried a maximum of 100 lbs (33 kgr). The animals served only between certain mansiones and had to be exchanged at certain determined spots. Their hay was to be exacted from the local population. The mansiones also served as central offices for all taxes paid "in natura".

The cursus velox could only be used by high officials, their families and the imperial messengers (tabellarii) (131). Originally this service was controlled by the praefectus praetoriae but later a special official vir illustris magister officiorum was appointed. The maximum loads of the different types of vehicles was carefully prescribed to avoid excessive wear of the roads (132). The average speed of the imperial post was some fifty Roman miles a day (5 miles per hour) (75 km). Special despatch bearers did sometimes attain double that amount, but 240 km a day allowing for halts is the maximum ever reported (133). It was certainly already better than the 24 miles which Cicero could do in 51 B.C. as a provincial governor (134). The regularity, certainty and speed of travel of the cursus publicus were not surpassed until the days of Napoleon. The ox-drawn carts travelled some 10—15 km a day.

The Romans tried to reduce the time of land transport and to increase its capacity in order to overcome the difficulty of feeding the great cities. The solution adopted was a control of most of the land and sea transport by liturgical (compulsory) services. Stables were built by forced labour, one quarter of the horses needed for the mansiones were requisitioned yearly from the population living nearby. The excellent effort broke down even when economy dominated road building during the later Empire. Transport costs, which doubled the price of grain for every 100 miles of transport, were so high as to make the carriage of wares, whose value was small in relation to bulk or weight, over any appreciable distance out of the question. Corn, ores or metals went straight to the nearest harbour and then by water. Thus local traffic may have been dense but long-distance transport was negligible. The development of roads helped only in a few cases, e.g. it brought down the price of vegetables in Rome.

The great and apparently beneficial road-system broke down through official abuse and the fundamental mistake in providing for its maintenance by compulsory exactions. There was an increasing and often fraudulent demand for free transportation of military and state officials. The municipial authorities already loaded with work could not exercize a proper control. Drastic reforms by Julian, Valentinian I and Theodosius helped only temporarily. The imperial fiscal administration had by the time of Constantine already destroyed one by one all the chief sources of revenue and traffic by land and sea gradually dwindled to a fraction of its former volume as private enterprise was completely absorbed by the state.

Having already mentioned milestones and directionposts as aids to travellers we should now devote a few words to maps and travelling handbooks. The Babylonians and their ancestors had had their maps (135) and the Greeks had greatly improved this art (136). Cartography was a well-known art at the beginning of our era. Maps of all types were available (137). During the consulate of Caesar and Mark Antony a survey of the empire was started by Zenodoxus, Theodotus and Polyclitus and finished in 25 years. Augustus' friend M. Vipsanius Agrippa ordered a new survey of Spain, Gaul (including the Low Countries) and the Danube provinces. The results were embodied in a sculptured marble map hung near the Pantheon at Rome. A 40′ × 60′ marble map of the 14 quarters of Rome was ordered by Vespasian, parts of this "Forma Urbis" have survived.

There were also maps for the general staff which accurately depicted the roads, the mansiones and their distances (138). A fragmentary copy of such a map on the leather shield of a soldier of Dura Europos gave the route and mansiones on the Black Sea highway from the mouths of the Danube to Artaxata in Armenia (139). It is the prototype of later "itineraria" like that of Ravenna (VIIth century A.D.) and the Peutinger map made by Castorius, of which we possess a copy of the XIIth century. This latter map is mathematically deformed to suit the pocket in the form of a roll, but all the distances are marked correctly (140).

Part of the troubles of ancient land transport sprang from the inefficient use of draught-animals (141). This was first of all due to the ancient form of harness, which was not a shoulder-collar with breast-strap, but which throttled a horse or mule that pulled too hard. Each animal could pull a maximum of 62 kgr, only a quarter of the modern figure. Also the only way of yoking more than one animal to a cart consisted in augmenting the amount of yokes. The Romans did not know how to arrange them in line tandem and thus not only complicated their manipulation (by multiplying the number of reins) but also reduced the total energy-output considerably.

Again horses were not shoed in Antiquity (142). Though the problem is still far from settled it seems that the ancients knew metal, leather or straw soleae, hypodemata or hipposandales attached to the legs of horses, mules or camels if the ground was hard or slippery. Real iron horseshoes, attached permanently with nails, seem to have been adopted from the nomads by German, Aleman, Helvetian & Sueban tribes of prehistoric Europe about the second century B.C. In the northern Roman provinces a few horseshoes were found before they become common about the eighth century A.D. Then a type of horsehoe with a better grip on the hoof were generally used from which type the different local European types began to evolve in the ninth century, when the emperor Leo mentions them in his Tactics, together with the rein and the stirrup both unknown to the Romans (143), which made riding a horse more difficult and which hindered the evolution of cavalry as a heavy arm.

Stirrups and horseshoes contributed to the rise of cavalry as an important "shock-tactic" element in warfare and they formed one of the cornerstones of the rise of chivalry, the mounted gentleman and his retainers who rode the highways in Chaucer's days along with the travelling clergy and merchants.

Roads and traffic in the Middle Ages

The fall of the Roman Empire meant the eclipse of central power. Hence systematic organised road-building and repair and all postal services centered on Rome were disrupted. Economic life was decentralised for several centuries and only after the Moslem armies from the south and the Viking inroads from the west and north were repulsed did western Europe recover sufficiently to start setting its house in order. A change was then slowly taking place, the western world was no longer Mediterranean, but Europe started to look westward, and towards the ocean after the discoveries of the fifteenth centuries. In the Byzantine Empire and that of Islam the traffic lanes did not change, the Roman and Hellenistic tracks were maintained and repaired. Ibn Khurdadhbieh in the ninth century could still describe a messenger-service with 930 relay-stations. Only slowly did this system disintegrate, though as early as the tenth century the last ancient bridge over the Tigris collapsed.

In the West the Roman roads survived longer than is usually supposed though by the end of the Middle Ages many were impassable, broken up by the farmers in need of building stones or gradually encroached upon by landowners. However for several centuries after the fall of the Empire they still remained excellent traffic lanes for a much diminished traffic. Several of them are mentioned in the early sagas (144). They remained the routes for the dispersion of ideas and manuscripts (145).

Even in these "dark ages" many towns, monasteries and officials had their messengers which carried first oral messages, later letters too. In many cases there were regular services of such "veredarii", "nuntii", and "missi" who had their passports and identification papers and in many ways ressemble their Roman predecessors. First these messengers travelled mostly on foot but by 1400 horses were commonly used. However we do not hear of any organised system of fire-signals or the like.

In the new decentralised world man was again reminded of the fact that traffic did not only depend on physical and geographical conditions but also, and now often more, on residential and trading facilities on the way, proper guarantees of safety and security and above all comparative freedom from imposts and taxes. Without central power money could not be raised in sufficient quantities to finance new construction and repairs. This lack of money greatly retarded the growth of a suitable new road-system with the proper orientation for medieval economic life.

Our data on medieval traffic, travel and road topography (147) are relatively scarce, but though in theory road-maintenance was an obligation of the land owners nothing came of this. Hence compulsory labour was used for roads and bridges with little succes, only in cities there was sufficient control. The tax raised for the repair of roads (viage) and bridges (pontage) helped in several cases to keep the traffic system in fairly decent repair. However medieval trade was more and more made to pay innumerable other tolls and imposts on roads, rivers, sea and markets. In England they were not so numerous as in France or Germany, though they tended to increase all over medieval Europe. In Germany some of the great Hanseatic routes were comparatively free, but in the rest of the country these tolls gradually succeeded in clogging internal trade, at the end of the Middle Ages. There were 30 tolls on the Weser by 1300, 35 on the Elbe, 30 on the Main and 80 on the Danube in Austria alone. Their number on the Rhine rose from 19 (1200 A.D.) to 35 (1300), 50 (1400) and finally 60 (1500 A.D.). The main weight fell on local traffic, reinforcing particularism and the self-sufficiency of local economies. They suppressed the rise of industry and agriculture. Hence the so-called "Reformation of King Sigismund" (about 1450) pleads free communication throughout the empire with just sufficient tolls to repair the roads and bridges.

Still apart from the protests against being fined for not repairing roads land owners did on occasions petition for "turnpikes" and their tolls. Then on many occasions money was given by municipalities and private individuals for the specific purpose of paving or repairing a certain road. For the building of roads and bridges belonged to the "trinoda necessitas" (threefold public duties). Thus the "congrégation hospitaliers des frères pontifes" built many bridges in twelfth-century France. In 1315 pious charity paid for the repair of roads near Durham. Monasteries, fraternities and guilds were very active, but in the thirteenth century the crown often took the initiative. The "strata publica" was part of the legal doctrine of thirteenth century France and in English Common Law the "king's highway" not only meant military roads but also all highways leading to towns and markets.

In this part of Europe, even in Low Countries, the drainage of the old Roman road was fairly good and hence the system survived more intact there. In Spain (VIIth century) the old roads were still used though tolls were heavy, there was some state control. Repairs of roads were perhaps too few, but we constantly find such data throughout the centuries. Thus in 1353 Edward III ordered paving the road to Westminster which had been paved in 1314 (148). Many of these roads were the old Roman roads, but in more cases deviations or new tracks sprang up to dodge tolls and taxes. Also medieval traffic had other demands. There were the great pilgrimages to Rome, Compostella and other famous shrines. New markets and fairs demanded different traffic lines (149). To Compostella alone four big routes led from France. In many cases two or more new routes sprang up besides the old Roman road betweens uch towns as Paris and Lyons. Laws often try to force men to work only on bridge repairs where they were "of old".

Merchandise went in wheelbarrows or on porters' backs over the Alps from the Rhine to the Po valley in the eleventh century. During the next century trade choose the route through the Rhône valley and then to Genoa though it was thrice as long. In the thirteenth century the Alpine passes were repaved and improved and trade went back to its old routes. Sometimes the Roman road had become a mud-track and the new route was preferred. The concept of the ideal road was changing too with the dominant means of transport. The Roman road was in reality an ideal marching road and its vehicles played a less important part because of several reasons already discussed. Now gradually a highway of cobbles or broken stone on a loose foundation of sand, which could expand with heat and cold, and which was easily repaired, became the dominant type.

The medieval state built "chemin ferré" with blocks cemented with mortar (a sand-lime-river mud mixture) on a sand foundation. Then there were the gravel or broken stone roads on a sand or earth base akin to the ancient rammed "lithostrata". Finally there were the cobble-roads such as that between Verberie and Senlus (1322) which the town of Ghent paid to promote its Paris trade (150), or the fifteenth century Bavai-Tournai road illustrated in the Chroniques de Hainaut of Jacques de Guise. Incidentally these French and Flemish paviours seem to have sat on four-legged stools and have moved forward as they laid the cobbles. The German paviours had the well-known one-legged stool and moved backwards as they laid their pavement.

The early Middle Ages saw the "colliers", the medieval coolies carrying or wheeling their loads along the road. The bulk of the goods was transported in two-wheeled carts ("bronnette") or more still in the four-wheeled "car" or "carrette", a flat-boarded peasant cart which carried about three times the load of the bronnette. The main alternative was not the packhorse but the barge or vessel. Such goods travelled about 22—35 km a day in flat country because of the better horse-collars which now made the use of the faster horse possible and hence longer journeys (151). Thus even with worse roads road-transport need not be more expensive. The transport of wool from the Cotswolds to Calais increased its price with 40%. In the thirteenth century the price of wool rose with 1.5%, that of grain with 15% for every 50 miles of transport. Of course this compares unfavourably with sea-traffic where Gascon wine was brought to Hull or Ireland at the cost of only 10% of the f.o.b. price. In fact the transport service was sufficiently efficient to show a lower ratio of trading costs to total costs during the Middle Ages than at present. It suffered most from political instability.

Messages and travellers went by foot or on horseback. Travelling by cart was still frowned upon, the term "karrenritter" was an insult. Travel in flat country about 50 km, in mountains some 40 km a day was quite common, hence not much more than in Antiquity. Though the lodging houses of the cursus publicus had disappeared, private houses, inns and xenodochia (pilgrim guesthouses established by religious communities) served the travellers. It has been calculated that most of the highways had to carry an average 1000 Tons a year though the salt-roads near Münich show a 7000 Tons salt-transport on three roads in 1370.

City streets and urban traffic

The central authority so essential for the development of a road system was present in the earliest cities from the earliest times. Hence in the cities of the Nile, Euphrates, Tigris and Indus valleys care was often given to the paving of streets, drainage and lighting and general for traffic and hygienic conditions of the community.

The Indus valley cities e.g. Mohenjo Daro had exemplary straight-angle city-plans, a definite authority must have prevented the rise of such tortuous alleys charactcristic of many later cities. The houses had a pronounced batter and never encroached on the streets like they do in a modern bazaar. All houses had latrines and bathrooms disposing their waste water into street drains. Houses also had rubbish shutes at the foot of which there were sometimes rubbish bins at street level, such as also stood at convenient places in the streets. The waste water entered the street drains via tightly closed brick-lined pits, with outlets to the street drains at about three quarters above the bottom. They seem the ancestors of our septic tanks and grit chambers.

Each street or lane had one or two drainage channels about 18—24″ below the street level and 9 × 12″ or 18 × 24″ in area, covered with stone slabs or corbelled roofs. The streets were usually aligned from east to west or south to north. Some of the important streets were

15′—33′ wide but the average is 9—12′. Most were unpaved though some herring-bone brick pavements were found in which the joint were grouted with gypsum-mortar or bitumen. In later levels (800 B.C.) pounded mixes of clay and potsherds made good floors and pavements (152). In later levels there was much less care for these details.

In Mauryan India (300 B.C.—150 B.C.) conditions were less ideal, though the ancient texts propagate ideal city-plans and wide streets (153). Excavations at Bhîtā and Taxila (154) proved that these towns had 10—12′ streets with a 1—2′ ditch in the middle. The basements of the houses could be entered from the inside alone (as already reported by classical authors) and "the town is divided by narrow irregular streets just like Athens" (155). We find little of the slab-pavement or crushed stone (gravel) surface, "which they had to repair every year" nor of the "drains on both sides of the road for the passage of water" (156).

In ancient Mesopotamia there was, from the earliest times, a tendency to build square or rectangular blocks of houses. The streets and lanes were mostly unpaved but many had sewers connected with laterals with the water-flushed latrines in the houses, e.g. at Eshnunna. Most streets were rather narrow and most of the ordinary refuse was cast into them. Though floors of bricks in bituminous mastic were known at a fairly early date (2000 B.C.), streets with stone-slab pavements are rare. Such efforts were mainly confined to the processional roads already described. During the Assyrian and Neobabylonian Empires such pavements became more common and more attention was given to the lay-out and appearance of the city (157). Similar conditions were found in other countries in the Ancient Near East (158) and even of ancient Egypt, though in the latter country very few entire towns have been excavated (Tell Amarna, Kahun).

Though drainage and sewerage of the Minoan palace of Knossos (1900—1700 B.C.) is excellent we are not informed about the settlements of the people (159). Conditions were far less ideal in classical Greece. The polis generally evolved without systematic town-planning, which was introduced by Hippodamos of Miletus (450 B.C.) according to Aristotle (160). Such town-planning came into effect in Hellenistic days (161). Spacious street-planning belongs to the same period. The main street of Alexandria was 30 m wide, the by-streets were 6—7 m. In the older towns they were seldom more than 4—6 m. Some processional roads like those in Miletus were about 20 m wide. The big road from Athens to Peiraeus was 14—15 m wide but the main road from the agora to the Acropolis was only 5 m. In general city-streets were narrow, dirty and unpaved. Thucydides writes of the "muddy village streets" of Sparta (162) and the agora of Elis was unpaved for it was used for the training of horses (163). The Theban army got lost and was defeated in the darkness and the mud of the streets of Plataea.

In some cases these streets were paved with slabs or with chips rammed into the subsoil (and sometimes grouted with lime mortar). The latter type was called "lithostratos". A "macadam" surface of this type required good drainage. According to STRABO drainage was unknown before the Roman rule (164). The older "macadam" streets must have suffered not only from this bad drainage but also from the refuse which was flung into the street with the outcry: "Existo" (stand out of the way) (165). ARISTOPHANES also complains about the muddy conditions of the Athenian streets (166). The authorities had to fight against these unhygienic conditions (167). Hence the astynomoi who controlled building in Athens also had to supervise the cleaning of drains and the sweeping of streets (168) which was executed by the koprologoi. The edict of 320 B.C. formally forbade throwing refuse into the streets and refuse-heaps have been found outside several Greek towns. At the same time the building of balconies and encroaching on the street made one liable to fines (169). The work of the astynomoi was later confided to the agoronomoi, who were responsible of all matters concerning the "agora" or market.

The drainage of lanes and streets was confined to one or two ditches along the side of the houses or slab-covered channels running irregularly through and across the streets. Sidewalks were non-existant except in the larger Hellenistic towns. Carriages and cars were mostly forbidden to enter the cities. The narrow streets, which suited the views of ancient experts proclaiming that the wind brought humidity and diseases (170), were rarely lighted. Some towns like Athens had a police-force of slaves (in Athens 300, later increased to 1000) called "epimeletai" which were also supposed to control the traffic (171). The maintenance of these streets was the duty of the board of "hodopoi" (172) who obtained money for this purpose from the Treasury or from rich private individuals (173).

In the older Italian towns paved streets are mentioned from 400 B.C. onwards. In Rome itself no pavement was found to antedate the Gallic invasion (390 B.C.). When Rome was rebuilt after the fire the square block plan of the Iron Age Terramare settlements with its cross of two main streets (cardo and decumanus) was adopted. The new colonies and cities generally adhered to similar town-plans, which were rigidly adopted for all army camps and forts. The forum of Rome was first paved with rectangular slabs (174). Then the street from the temple of Mars to the Capuan Gate was paved with slabs (saxo quadrato) of "peperino" about 296 B.C. (175). Next was the street from the temple of Mars to the Campus Martius about 190 B.C. (176).

Streets in older Roman towns were narrow according to our standards, usually not wider than 5 m except main streets (7 m and over). The emperor Augustus limited the height of buildings to 70′, which maximum was reduced to 60′ by Trajan. Thus building was limited to five stories and light could still enter the streets. Wheeled traffic was

Fig. 38.
A camel of the 7th century B.C. Arabian archers on single-humped camels withstand the infantry of Assurbanipal

forbidden during the day except on such public occasions as religious festivals or games. The transport of building materials (except for public buildings) and the like had to take place between dusk and dawn. Caesar had to repeat this order in 47 B.C. and so did Claudius, Marcus Aurelius and other emperors (177). Refuse and rubble had to be transported at night. Driving was considered ungentlemanlike, one walked, rode a mule or was carried in a sedan-chair. Sidewalks were common and blocked the house-entrances from cars driving into the inner court of the houses. The narrow streets were considered hygienic in Rome too.

By 174 B.C. paved streets were common in Rome and basalt slabs were generally used. About half the width was taken up by the sidewalks. About the same time streets in provincial towns such as Alatri or Puteoli were paved too (178). We know much about the history of pavements in certain fully excavated towns like Pompeii (179). In this town streets with a "macadam" surface were rather wide upto 200 B.C. For the next century the adoption of the peristyle house led to the formation of narrow arcaded streets which were typical of many cities of the Imperial period (Timgad, Lambaesa, Dugga). Slab pavement was now in regular use. The inhabitants of the street had to pay their share in the costs of repairs one month after the aediles had finished the job. Finally from 50 B.C. onwards the houses encroached still more on the, streets and houseowners possessing adjacent blocks often closed them to traffic by chains or stepping stones making them private lanes.

In general the towns began to pave their streets with more fervour after Caesr's law of 47 B.C., which also stipulated that refuse should no longer be thrown into the streets, these would now be cleaned with water regularly by the workmen of the tribuni rerum intentium. Augustus fixed the width of the decumanus at 40′, the cardo at 20′, side-streets (vicinae) at 12′ and lanes at a maximum of 8′. This is in line with the dimensions found at Pompeii. The sidewalks were held by curbstones (30—45 cm wide) interrupted at regular intervals by drains leading to the sewers (cloacae).

In ancient Rome such sewers for the surface draining of streets were adopted from the Etruscans. The Cloaca Maxima (14′ high and 11′ across) was built about 500 B.C. to drain the area of the Forum (180). The three little streams which formed the natural drains of Rome (Petronia annis, Spinon, Nodinus) were gradually converted into cloacae, which system was extended about 300 B.C to make Pliny proudly speak of the "floating city of Rome" (181). The foot- and shower-baths and the wash-stands in many Roman houses gave immediately into the sewers as did the latrines present in every Roman house. There were also public lavatories in many Roman cities, one for 45 persons in Puteoli, for 28 persons in Timgad, and Rome itself about 315 A.D. had 144 "latrinae publicae" (with water flushing) and 116 "necessariae".

At the street corners stepping stones allowed foot-passengers to cross the streets with dry feet. By enlarging these stones one could turn an alley into a private lane. Spur posts were a regular feature of Roman towns. In the Imperial period pavements were not only financed by the state (182) but also donated by private individuals (183). Costs appear rarely in our inscriptions, but the town-council of Nuceria had 1165′ of street paved at the cost of 22½ sesterces per foot, the money of which was obtained from tolls (184). Usually a "street-tax" was collected, and though usually special paviours were employed a statute of Urso would suggest that formerly forced labour had been responsible for paving. The rôle of the astynomoi in the East was taken over by the aediles in Roman towns, who in the Imperial Period made way for special curations viae by law of Augustus (22 B.C.).

Antiochia was the first city in the world to ligh' its streets with tarred torches about 450 A.D. (185). Lanterns were well-known in Antiquity, cylindrical and square prismatic lanterns with a conical top were quite common. Thin plates of horn or mica and later glass often formed the windows. The "lanternarius" was the slave who preceeded his master on his evening walks. Statuettes of such slaves with their lanterns are quite common too. These lanterns were made of metal, wood and pottery (186). Besides there were many forms of oil-lamps with wicks and candles used in such lanterns. Portable lanterns were made in great quantities at Capua since the first century B.C.

Hence before the days of Antiochia the main streets of Roman towns must have been quite properly illuminated by the lamps in the shops and over the house-entrances. In the Via dell' Abundanza (Pompeii) there were at least 285 lamps on counters or above doors on a stretch of some 500 m. "Second street" in the same town had 396 lamps in 132 shops over 576 m and the 700 m "High Street to Stabiae" had some 500 lamps (187). Then there were the lamps on the corners of streets under the statues of gods or house and family by spirits (lares compitales) (188) about which custom Tertullian complains (189) and which were finally forbidden by Theodosius the Great (190). Then again many temples (191) and often graves (192) were brightly illuminated at night and traffic at night may not have been as uncomfortable as generally supposed. By the sixth century A.D. street lighting was quite common in the Orient.

In the Middle Ages conditions were certainly worse, though our information is unfortunately too scarce. Most of the towns of western Europe were new creations starting without proper planning and little comfort such as paving or lighting. In German towns (193) paving started in the twelfth century very slowly and it became common practice only much later (Lübeck 1310, Nuremberg 1368, Berne & Frankfort 1339, Regensburg 1400). The smaller towns followed suit. Earlier streets were sometimes paved with "macadam" called "heisser Steinweg" or "teueres Pflaster" (expensive pavement). Then slabs, cobbles or bricks came into use.

Refuse was thrown into the streets and though the inhabitants were held to clean it the law could not always be enforced. When princes visited the town or processions were held there was a general clean-up. In some towns the law was regularly enforced (Göttingen 1330, Mechelen 1348) until it became a public service towards the end of the Middle Ages (Nüremberg 1490). Pigs were held in the streets and epidemics raged furiously. Drainage was seldom adequate though little brooks were turned into the streets to scavenge them. Lights were scarce except on bridges or in the case of fires. At night many streets were barred with chains.

In ALEXANDER NECKAM's days (194) the streets of Paris were in disrepair and full of mud. Little remained of the Roman pavement and indeed doggerels derived the name "Lutetia" from "lutum" (mud). The sewer was an open channel in the middle of the road into which refuse was thrown, in which pigs walked and into which the latrines often gave (195). Shoes had to have heavy, high and thick soles. As the streets were not lit one had to walk with a boy carrying a torch or a lantern. There was some light from the houses as lamps were lit by sic in the evening. Carts were not allowed in town, but horses were, they had by now displaced the ox as draft animals. Riding in carts was reserved for women or sick people.

In England conditions were hardly better (196). In Cambridge the king commanded the Major and Bailiffs to repair the pavement of streets and lanes and to compel everyone to pave the street before his tenement (1335-6). Such repairs were paid by "pavage" levied on many classes of goods. Northampton was granted such a levy by Edward I (1284) and other towns followed (Cambridge 1289, Nortwich 1277, Chester 1279, Liverpool 1329). This "pavage" was based on the weight of goods and the type of vehicle. Sometimes public spirited citizens left money for this purpose and thus 36 roads in or near London were paved in the period of 1358—1509.

In London each Alderman appointed "four reputable men" to repair and clean the streets (1280). They were the ancestors of the later "scawageours" or scavengers (1364) controlled by a "surveyor of streets" (1390). The actual paving, first done by the citizens, became an expert job extented in the fourteenth century under a "stone master". Southampton had an expert "town paviour" in 1482. The paviours of London were organised in 1479. They earned respectable wages fixed by the London ordinance of 1301. Several of them rose to some dignity and affluence, like the paviour and scavenger who became members of the Nüremberger "Zwölfbrüder" society.

These paviours used cobbles and sand for the markets and squares and gravel and sand for the lanes. Potholes were often simply filled up with faggots of broom and chips. In paving the old surface was rarely cleared away and this of course tended to raise the whole street. Pavements were properly compacted with hand-rams before use. The greatest wear of the streets came from iron-shod cart-wheels. When iron tyres were forbidden the wheels were protected with nails and promptly the use of all but flat nails is forbidden. Wheel-barrows (known at least since the thirteenth century) (197), "drags" and "sleds" completed the range of vehicles.

There was much encroaching on the street by stalls and by overhanging balconies. Hence a mounted man with a standard lance across his saddle would now and then ride in medieval cities to test this, and fine all the evildoers. Heavy traffic was compelled by law to take certain streets and gates, but these regulations had no avail on the long run and had to be repeated constantly. Traffic congestions at the toll-gates were frequent and so were accidents and law-suits about rapid or careless driving. Rubbish and filth in the streets accumulated and the pigs running about did not improve these conditions. Hence the citizens are often ordered to clean the frontage of their house once a week (usually on Saturday) and cart the rubbish to the dunghills outside the city walls or dispose of it in rivers. It took several centuries to restore conditions to the level of Antiquity.

Bibliography

1. Navarro, J. M. de, *Prehistoric Routes between Northern Europe and Italy* (*Geogr. J.* vol. 66, 1925, 481); Clark, J. D. G., *Prehistoric Europe* (London, 1952); Déchelette, Fr., *Les voies de pénétration de la civilisation dans la Gaule celtique et romaine* (*Rev. Géogr.* Lyon XXXI, 1956, 109—113
2. Landsberger, B., *Assyrische Handelskolonien in Klein Asien* (Leipzig, 1925); Heyman, R., *Transportation in the ancient Middle East* (*Koroth* (Israël), 2, 1958, 1/2, 76—77)
3. Drioton, E. & Vandier, J., *L'Egypte* (Paris, 1938)
4. Salonen, A., *Nautica Babyloniaca* (Helsinki, 1942); Salonen, A., *Die Landfahrzeuge des alten Mesopotamiens* (Helsinki, 1951); Salonen, A., *Hippologica Accadica* (Helsinki, 1955)
5. Blum, O., *Die Entwicklung des Verkehrs* (Berlin, 1941); Forbes, R. J., *Notes on the history of ancient roads and their construction* (Amsterd., 1934) (photost. reprint, Amsterdam, 1964); F. Freiherr von Richthofen, *Siedlungs- und Verkehrsgeographie* (Berlin, 1908); Birk, A., *Die Strasse* (Karslbad, 1934); Hettner, A., *Die geographische Verbreitung der Transportmittel des Landverkehrs* (*Z. Ges. Erdkunde* vol. 29, 1894, 271—289); Mitman, Carl W., *An outline development of highway travel* (*Smithsonian Report for* 1934, 1936, 325—345)
6. Semple, E. C., *Geography of the Mediterranean Region* (London, 1932); Cary, M., *Geographical Background to Greek & Roman History* (Oxford, 1947); Sarton, G., *Unity and Diversity of the Mediterranean World* (*Osiris* vol. 2, 1936, 406—463); Lee, N. E., *Travel and transport through the ages* (Cambridge, 1955); Wiesner, J., *Vor- und Frühzeit der Mittelmeerländer* I & II (Leipzig, 1943)
7. Hippesley Cox, R., *The Green Roads of England*; Clay, R. C. C., *Some prehistoric ways* (*Antiquity*, vol. I, 1927, 54); Peake, H., *Prehistoric Roads* (*Archaeol. Cambr.* Ser. 6, vol. 17, 1917, 353); Watkins, A.. *The Old Straight Track* (London, 1933); Hutton, J. H., *Assam Megaliths* (*Antiquity* vol. 3, 1929, 335); Haldane, A. R. B., *The drove roads of Scotland* (*Geography* G.B., 1955, 40, no. 190, 292)
8. Forbes, R. J., *Notes on the history of ancient roads and their construction* (Amsterdam, 1964, chapter V); Louis, R., *Construction en bois à Yevrès (Cher)* (*BSAF*, 1950—1951, 31); Goodwin, H., *Prehistoric wooden trackways of the Somerset levels* (*Proc. preh. Soc. G.B.*, 26, 1960, 1—36)
9. Evans, E. M. P., *Maltese cart-ruts* (*Antiquity* vol. 8, 1934, 339—342); Zammit, T., *Prehistoric Carttracks in Malta* (*Antiquity* vol. 2, 1928, 18); Zammit, T., *Prehistoric Malta* (London, 1932); Evans, J. D., *Malta* (London, 1959, *App.* I, p. 189—191)
10. Sir Evans, A., *The Palace of Minos* (vols. II^1 & II^2), (London, 1921—1936)

11. KOLDEWEY, R., *Die Pflastersteine von Aibur Shabu in Babylon* (*Wiss. Veroff. Dtsch. Orient Ges.* No. 2, Berlin, 1901); KOLDEWEY, R., *Das wiederentstehende Babylon* (Leipzig, 1913, 23); see also *Studies in Ancient Technology*, vol. I (1964) page 80
12. ANDRAE, W., *Alte Feststrassen im Nahen Orient* (Leipzig, 1941)
13. HUNGER, J., *Heerwesen und Kriegsführung der Assyrer* (*Alte Orient* vol. 12, no. 4, 1911)
14. LUCKENBILL, D. D., *Ancient Records of Assyria* (Chicago, 1926, vol. I, 75)
15. Jer. VI.1; Judges XX.38; TORCZYNER, H., *The Lachish Letters* (London, 1938, 75)
16. DOSSIN, G., *Signes lumineux au pays de Mari* (*R. Assyr.* vol. 35, 1938, 174—186); FORBES, R. J., *Studies in Ancient Technology*, vol. VI (Leiden, 1958, 168 ff)
17. *VAB* I, 102; XI, 26; *VAB* VII, 264; III.10
18. *Z.f. Assyriologie* vol. 15, 238 & 35, 217
19. WINKLER, H., *Altorientalische Forschungen* I, 298, I, 8; TALLQVIST, K. L., *Die assyrische Beschwërungserie Maqlu*, III.147
20. HERODOTUS, I.186; BUREN, A. W. VAN & LAMMERT, F., *Pons, Brückenbau, Pons im Kriegswesen, RE Erste Reihe* XXI, 2, 1952, 2428—2484; HOLLAND, L. A., *Forerunners and revivals of the primitive Roman bridge* (*Proc. Amer. Phil. Assoc.* 80, 1949, 281—319); SANTARELLA, L., *Arte e tecnica nella evoluzione dei ponti* (*Atti richerche studi* XIII, Milano, 1933); STEINMAN, D. B. & WATSON, S. R., *Bridges and their builders* (Dover Publ. T 431, New York, 1957)
21. UNGNAD, A., *Babylonische Briefe*, no. 15 (Berlin, 1914)
21a. ANATI, E., *Bronze Age chariots from Europe* (*Proc. Preh. Soc. G.B.* 1960, 26, 50—63); MUSTAPHA EL-AMIR, *note on thyr. r in boundaries of Ptolemaic houses at Thebes* (*ASAE* 53, 1955, 135—138); BOTTI, G., *Il carro del sogno* (*Aegyptus* 31, 1951, 192—198); BRUNNHUBER, G., *der vierräderige Wagen in Vorzeit und Antike* (Wien, 1952); CALLAGHAN, R. T., *New light on the Maryannu as chariot warriors* (*Jb. Kleinas. Forschung* I, 1950/51, 309—324); GORDON CHILDE, V., *The first waggons and Carts from Tigris to the Severn* (*Paper* no. 6 *Proc. Preh. Soc.* 1951, part 2); FOLTINY, S., *The oldest representations of Wheeled Vehicles in Central and Southeastern Europe* (*AJA* 63, 1959, 1, 53—58); HAUDRICOURT, A. G., *L'origine de la duga* (*Ann. Hist. Soc.* II, 1940, 34); GIULIETTI, L., *Of some ancient wheels discovered in Italy* (*La rivista RIV* no. 11, 1961, 3—17); SCHACHMEYER, F., *Streitwagen und Streitwagenbild im alten Orient und bei den mykenischen Griechen* (Anthropos 46, 1951, 705—753); VAN DER WAALS, J. D., *Prehistoric disc wheels in the Netherlands* (Groningen, 1964)
22. LAUFER, B., *J. Race Development* vol. 5, 1914, no. 2
23. REISNER, G., *Tempelurkunden aus Tello*, no. 129, IV, 9; J. WIESNER,, *Fahren und Reiten in Alteuropa und im alten Orient* (*Der alte Orient* vol 38, 1939, 2/4); SCHULMAN, A. R., *Egyptian representations of Horsemen and Riding in the New Kingdom* (*JNES* 16, 1957, 263—271); STREUBEL, FR. *Fahren und Reiten im Uebergang vom Altertum zum Mittelalter* (Hamburg, 1954); SALONEN, A., *Hippologia Accadica* (Helsinki, 1955)

24. Xenophon, *Cyropaedia* VIII.6.17; Herodotus, VIII.98
25. Herodotus, V.52
26. Herodotus IX.3; Aristotle, *De mundo* VI, 11—12
27. Diels, C. H., *Antike Technik* (Leipzig, 1924, 77); Chapot, V., article "Signum" in Daremberg-Saglio, *Dictionnaire* 1334—1335
28. Esther 8.10; 3,13
29. Plutarch, *Alexander* 18
30. Calder, *The Royal Road in Herodotus* (*Class.* Rev. vol. 39, 1925, 7)
31. Diodorus, II.1
32. Diodorus, I.2; Xenophon, *Anabasis* I.2; 1.5; III.5; IV.1; V.1—2
33. Xenophon, *Cyropaedia* VI, 2.25
34. Forbes, R. J., *The coming of the camel* (see chapter IV)
35. Phelps Grant, Chr., *The Syrian desert* (London, 1937); Will, E., *Marchands et chefs de caravane à Palmyre* (Syria XXIV, 1957, 262—277)
36. Ammianus Marcellinus, Book XIV,4 (5)
37. Sir MacDonald, Geo, *Rome in the Middle East* (*Antiquity* vol. 8, 1934, 373—380); Bradford, J., *Ancient Landscapes* (London, 1957)
38. Strabo, XVI, cap. 748
39. Thucydides, I. 3.4; Salmon, P., *La population de la Grece antique* (*Bull Soc. Belge Géogr.* LXXIX, 1955, 34—61)
40. Thucydides, I.120.2
41. Strabo, V. cap. 235
42. Curtius, E., *Zur Geschichte des Wegebaus bei den Griechen* (*Abh. K. Akad. Wiss. Berlin*, 1854); Chapot, V., article "Via" in Daremberg-Saglio, *Dictionaire des Antiq. Gr. et Romains*
43. Steffen, *Karten von Mykenä* (Textband, 8)
44. *Odyssey* Z, 390
45. Pausanias, VIII, 54.5; X.32.8; X.5.2; II.15.1; X.32.6; II.38.4; X.35.8; II.11.3; I.44.6; Tréheux, J., *Une nouvelle voie thasienne* (*Bull. Cor. Hell.* LXXIX, 1955, 427—441); Young, J. H., *Greek roads in South Attica* (*Antiquity* XXX, 1956, 94—97)
45a. Burford, A., *Heavy transport in classical Antiquity* (*Econ. Hist. Rev.* XIII, 1960, 1—18)
46. Pausanias, II.1.7; VIII.26.1
47. Stobaeus, Flor. II.275; Aristophanes, *Ranae* V. 109. 112
48. Herodotus, VI.57
49. Strabo, VII.7.4; Pliny, *Nat. Hist.* VI.61; Aristotle, *Pol.* III.39.8
50. Valerius Maximus, III.7; Rochas, A. de, *Les rainures des chemins antiques* (*La Nature*, Aug. 26, 1907, 207)
51. Mure, *Journal of a Tour in Greece* vol. II, 251 (London, 1842); Caillemer, *Les voies à rainures chez les anciens* (*Congrès arch. de France* 1879, 277)
52. Euripides, *Ion* 756
53. Sophocles, *Oedipus Tyrannus* 733
54. *C.I.Gr.* 1688
55. Darmstaedter, E., *Feuer-telegraphie im Altertum* (*Umschau* vol. 28, 1924, 505—507); Fluss, V., *Die Feuerpost im Altertum* (*W. Bl. Freunde Antike* vol. 9, 1933, 77); Merriam, *Telegraphing among the ancients*

(*Pap. Archaeol. Instit. America*, Class. Ser. III.1, 1890); REINECKE, G., *Feuertelegraphie im griechischen Altertum* (*Archiv f. Post und Telegraphie* 1935, 143—145); See also my *Studies in Ancient Technology*, Vol. VI (1958), 168 ff.
56. THUCYDIDES, III.77.3
57. HERODOTUS, VII, 131
58. ARRIAN, I. 26
59. THUCYDIDES, II.98 & 102
60. Hibeh-Papyrus 110 (edit. GRENFELL and HUNT, London, 1906, Part I)
61. DIODORUS, XIX, 57, 5
62. PREISIGKE, FR., *Die ptolemaïsche Staatspost* (*Klio* vol. VII, 1907, 241—277)
63. CHARLESWORTH, M. P., *Trade Routes and Commerce of the Roman Empire* (Cambridge, 1924, 18)
64. MURRAY, G. W., *The Roman Roads and Stations in the Eastern Desert of Egypt* (*J.E.A.* vol. XI, 1925, 138—150); MEREDITH, D., *Inscriptions from the Berenice Road* (*CdE* XXIX, 1954, 281—287); PRÉAUX, CL., *Sur les communications de l'Ethiopie avec l'Egypte hellénistique* (*CdE*) 1952, no. 53, 257—281)
65. TAESCHNER, F., *Verkehrslage und Wegenetz Anatoliens* (*Petermann's Mitt.* vol. 72, 1926, 202—206); BALLANCE, M. H., *Roman Roads in Lycaonia* (*Anat. Stud. Gr. B.* VIII, 1958, 223—234); OSTEN, H. H. VON DER, *Anatolische Wege* (*Eranos* XLIX, 1951, 65—84); YOUNG, R. S., *The campaign of* 1955 *at Gordion* (*AJA* 60, 1956, 266
66. FLEURE, H. J., *The Geographical background of modern problems* (London, 1932, 59)
67. BERGIER, N., *Histoire des grands chemins de l'empire romain* (Paris, 1622)
68. ISIDORUS, *Originum...* Lib. XV, cap. ult., num. 5 & 6; DEUTERO-SERVIUS, *Notes on Aeneid* I. 422
69. MERCKEL, C., *Ingenieurstechnik im Altertum* (Berlin, 1899)
70. DIODORUS, XX, 44.5; POLYBIOS, I.75, 4—5; STRABO, XVIII. 3.18 cap. 835
71. RANDALL-MAC IVER, D., *Italy before the Romans* (Oxford, 1928)
72. RANDALL-MAC IVER, D., *The Etruscans* (Oxford, 1927)
73. MARTHA, *Histoire de l'art étrusque*, 253; DENNIS, G., *Cities and Cemetries of Etruria*, London
74. ANZIANI, *Les voies romains de l'Etrurie méridionale* (*Mél. Archéol. Hist. Ecole Franc. de Rome* Vol. 33, 1933, 190); FELL, R. A. L., *Etruria and Rome*, Cambridge, 1924, 151; WARD PERKINS, J. B., *Etruscan engineering, road-building, watersupply and drainage* (Mélanges Grenier, Brussels, 1962, 1636—1643)
75. LIVY, V. 40
76. DIODORUS, XX, 44.5; POLYBOIS, I.75, 4—5; STRABO, XVIII.
77. LIVY, III.6; II.39
78. LIVY, VII.9; PLINY, *Nat. Hist.* 31.89
79. LIVY, X.47; FUSTIER, P., *Notes sur la construction des voies romaines en Italie. II Via Appia* (*REA* LXII, 1960, 95—99).
80. LIVY, XLI, 27
81. STRABO, cap. 322, 380, 391; EDSON, CH., *The location of Cellae and the*

route of the Via Egnetia in Western Macedonia (*C. Ph.* XLVI, 1951, 1—16); MAKARONAS, CH. I., *Via Egnetia and Thessalonike* (*Studies Robinson, I, St. Louis*, 1951, 380—388)
82. *C.I.L.* III.6123; *C.I.L.* III.7613, 7615
83. SICKLE, C. E. VON, *The repair of the roads in Spain under the Roman Empier* (*Class. Philology* vol. 24, 1929, 77)
83a. MARGARY, I. D., *Roman Roads in Britain* (London, 1957, 2 vols.); CRAWFORD, O. G. S., *Archaeology in the Field* (London, 1953, 60—86)
84. SICKLE, C. E., *The public works of Africa in the reign of Diocletian* (*Class. Philology* vol. 25, 1930, 173—179); CARPENTER, RHYS, *A Trans-Saharan Caravan Route in Herodotus* (*AJA* 60, 1956, 3, 231—242); LECLANT, J., *Per Africae sitientia, Témoignage des sources classiques sur* SALAMA, P., *Les voies romaines de l'Afrique du Nord* (Alger, 1951); EUZENNAT, M., *Les voies romaines du Maroc dans l'itinéraire Antonin* (Mélanges Grénier, Bruxelles, 1962, 595—610)
84a. AVI-YONAH, M., *The development of the Roman roadsystem in Palestine* (Israel *Expl. J.*, I, 1950/51, 54—60); GOODCHILD, R. G., *The coastal road of Phoenicia and its Roman milestones* (Berytus IX, 1949, 91—127)
85. Sir MACDONALD, GEO, *Rome in the Middle East* (*Antiquity* vol. 8, 1934, 373—380); KUHL, G., *Römische Strassen...* (*Palästinajahrbuch* vol. 24, 1928, 113—140)
86. MILLER, R., *Itineraria Romana* (Stuttgart, 1916); CHARLESWORTH, M. P., *Trade routes and commerce of the Roman Empire* London, 1924); KRÜNITZ, E., *Vom Strassenbau* (Berlin, 1794)
87. CARTELLIERI, A., *Die römischen Alpenstrassen* (*Philologus* Suppl. band XVIII.1, 1926); GRENIER, A., *Les voies romaines en Gaule* (*Mél. Arch. Hist. Ecole Franc. Rome* vol. 53, 1936, 2—24); FORBES, R. J., *Bibliographia Antiqua* (Leiden 1950—1963, vol. X & Suppl. I & II)
87. VITRUV, V. 9.7
88. VITRUV, VIII.1
89. PLINY, *Nat. Hist.* 36.60—70, 184—189
90. STATIUS, *Silvae* IV.3; FUSTIER, P., *Etude technique sur un texte de l'empereur Julien relatif à la construction des voies romanines* (Rev. Etud. Anc. Fr. 65, 1963, 1/2, 114—121); FUSTIER, P., *Nouvelle coupe de la voie romaine de Lyon à Roanna* (Rev. Arch. Est XIII, 1962, 124—127)
91. PALLADIO, *I quattro libri dell' Architectura* Lib. III, cap. III Venetia, 1570)
92. STUART JONES, H., *Companion to Roman History* (London, 1912, 52—66)
93. MATSCHOSS, C., *Staat und Technik* (Vortrag 52. Samml. V.D.I. Breslau, 1911)
94. BURR, W. H., *Ancient and Modern Engineering* (New York, 1903); MERCKEL, C., *Ingenieurstechnik im Altertum* (Leipzig, 1900); MILLER, R., *Itineraria romana* (Stuttgart, 1916); BIRK, A., *Die Strasse* (Karlsbad, 1934); Article "Via" in DAREMBERG-SAGLIO, *Dictionaire...*; LÉGER, A., *Les travaux publics aux temps des Romains* (Paris, 1875, 143—150)
96. FORBES, R. J., *History of Ancient Roads and their Construction* (Amsterdam, 1964, 160—165)
97. PLUTARCH, *G. Gracchus* 7

98. HIRSCHFELD, *Römische Meilensteine* (*Ber. Berl. Akad.* 1907, 165)
99. PLUTARCH, *G. Gracchus* 7; GALEN, *De Method Medendi* 9.8; KIELEY, E. R., *Surveying Instruments* (New York, 1947)
100. CODRINGTON, *Roman Roads in Britain* (London, 1918)
101. FERRAND, H., *Les ingénieurs militaires romains dans les Alpes* (*La Nature*, 6 mai 1905)
102. *C.I.L.* X. 6849
103. PAYNE, *On a Roman road near Rochester* (*Archaeol. Cantiana* vol. 23, 4)
104. MENGERINGSHAUSEN, J., *Die Trajanstrasse am unteren Donau* (*Technikgeschichte* vol. 22, 1933, 136)
105. CAESAR, *De Bell. Gallico* VIII. 14
106. TACITUS, *Annals* I. 61—63
107. ORT, J. A., *Oude wegen en landweren in Limburg*
108. TACITUS, *Annals* III.31
109. DIO CASSIUS, LIX.15.3
110. DIO CASSIUS, XLVII.17
111. *Codex Theodos Lib.* XV. tit. 3 & 6; Lib. XII, tit. 51; ULPIAN, *Ad Decretum on "Via Publicae"* I
112. *C.I.L.* X.5416; V.8668; V.1008a; X.3851; *Dig.* 31.10; PHILOSTRATUS, *Vit. Soph.* I.19
113. *C.I.L.* X.6954
114. *C.I.L.* III.218; III.471; JOSEPHUS, *Antiquities* XX. 9.7
115. *C.I.L.* VI.3824
116. *C.I.L.* IX.6075
117. HEITLAND, W. A., *Iterum* (London, 1925, 25); HOLLAND, J., ROSE, *The Mediterranean in the Ancient World* (London, 1934, 145—153)
118. KNAPP, CH., *Travel in ancient times as seen in Plautus and Terence* (*Class. Phil.* 1907, 21—24; 281—304); KNAPP, CH., *A note on travel in ancient times* (*Class. Weekly* vol. 28, 1935, 177—178); WELLS, B. W., *Trade and travel in the Roman Empire* (*Class. J.* vol. 19, 1923/24, 7—16, 67—78); HOHLWEIN, N., *Déplacements et tourisme dans l'Egypte Romaine* (*Chron. d'Egypte* No. 40, 1940, 253—278); KAPITAEN, G., *Schiffsfrachten antiker Baugesteine und Architekturunterteile vor den Küsten Ostsiziliens* (Klio XXXIX, 1961, 276—318)
119. *O.G.I.S.* 674; PEARL, O., *Transport charges in Egypt in the era of inflation* (*TAPhA* LXXXIII, 1952, 74—79); WOLFE, E. R., *Transportation in Augustan Egypt* (*TAPhA* LXXXIII, 1952, 80—99)
120. *Gnomon de l'Idiologue*, Berlin Museum papyrus
121. STRABO, XVII cap. 815; GESTER, W., *Beitrag zur Geschichte einiger Bezeichnungen für Gasthaus* (Vox Romana IX, 1946/47, 57—151); KLEBERG, T., *Hôtels, Restaurants et Cabarets dans l'Antiquité Romaine* (Uppsala, 1957)
122. Ostrakon Brit. Mus. 16467; Pap. London III. 1159; Pap. Florence 127
123. AMMIAN, MARCELLINUS, XXII. 16.18
124. DIO HALLICARNASSUS, III. 67; DIODOR, XX. 36
125. LIVY, XLII. 1
126. CICERO, *Ad Fam.* II. 17. 1
127. STRABO, V.4.13, cap. 251; CELLIUS, X. 13.9; RAMSAY, A. W., *A Roman*

postal service under the Republic (*J. Roman Studies* vol. 10, 1920, 79—86)
128. *C.I.L.* I. 550
129. Suetonius, *Div. Aug.* 64.3; Procopius, *Anecdota* c. 30
130. Holmberg, E. J., *Zur Geschichte des cursus publicus* (Uppsala, 1933); Gorce, D., *Les voyages... dans le monde chrétien des IVe et Ve siècles* (Paris, 1925); Pflaum, H. G., *Essai sur le cursus publicus sous le Haut Empire Romain* (Paris, 1940)
131. Desjardins, M., *Les tabellarii, porteurs des dépèches chez les Romains* (*BEHE* vol. 35, 1878, 51—81)
132. *Codex Theodos., de cursu publico* VII. 5—6—; VIII 1, 5, 7, 17, 16, 26, 28, 30, 47, 48, 364
133. Ramsay, A. W., *The speed of the Imperial Post* (*J. Roman Studies* vol. 15, 1925, 60—74); Elliot, C. W. J., *New evidence for the speed of the Roman imperial post* (Phoenix IX, 1955, 76—80)
134. Hunter, L. W., *Cicero's journey to his province of Cilicia in* 51 *B.C.* (*J. Roman Studies* vol. 3, 1913, 73—97)
135. Unger, E., *Ancient Babylonian Maps and Plans* (*Antiquity* vol. 9, 1935, 311—322); Meek, Th. J., *The Orientation of Babylonian Maps* (*Antiquity* vol. 10, 1936, 223—226); Gordon, C. H., *Points of the compass in the Nuzi tablets* (*R. Assyr.* vol. 31, 1934, 101—108)
136. Meissner, Br., *Babylonische und griechische Landkarten* (*Klio* vol. 19, 1923, 97—100); Heidel, W. A., *The frame of ancient Greeks maps* (New York, 1937); Thomson, J. O., *The History of Ancient Geography* (Cambridge, 1948)
137. Uhden, R., *Die Weltkarte des Isidorus von Sevilla* (*Mnemosyne* ser. III. vol. II, 1935/36, 1—28); Keune, J. B., *Reisehandbücher, Reiseuhren, Reise und Verkehr zur Romerzeit* (*Trierer Heimatblatt* 1922, I. 68—72, 94—98, 116—118, 148—150)
138. Vegetius, III.6
139. Cumont, M., *Un extrait d'une carte romaine d'état major* (*La géographie* vol. 43, 1925, 1—5); Uhden, R., *Bemerkungen zu dem römischen Kartenfragment von Dura Europos* (*Hermes* vol. 67, 1932, 117—125)
140. Freitag Drabbe, C. A. J. von, *Die Peutingerkarte in dem Lichte der neuesten Untersuchungen* (*C.R. Congres Int. Géogr.* Amsterdam 1938, vol. II. 209—210); Miller, R., *Itineraria Romana* (Stuttgart, 1916); Baschmakoff, A., *La synthèse des périples pontiques* (Paris, 1948); Brunsting, H., *Romeinsche mijlpalen in Nederland* (*Oudh. Med. Rijksm. Oudh.* XXVII, 1946, 28—36); Durand, D. B., *The Vienna/Klosternburg map corpus* (Leiden, 1952); Maury, J., & Percheron, R., *Itinéraires romains* (Paris, 1958); Richmond, I. A. & Crawford, O. G. S., *The British section of the Ravenna Cosmography* (*Archaeol.* XCIII, 1949); Stolte, B. H., *De Cosmographie van den Anonymus Ravennas* (Amsterdam, 1949)
141. Lefebre des Noëttes, *L'attelage, le cheval de selle à travers les âges* (Paris, 1931)
142. Kroon, H. M., *Hoefijzers uit de vroege Middeleeuwen* (*Tijdschr. Diergeneeskunde* vol. 53, 1926, 579—586); Michon, E., *La trouvaille faite à Eauze* (*BSAF* 1927, 107—109); Winkelmann, Fr., *Über das Hufeisen*

(*Germania* vol. 12, 1928, 135—143); KAUFMANN, D. B., *Horsehoes in Antiquity* (*Class. J.* vol. 26, 1930/31, 619—620; VAATH, J., *Römische Hufeisen* (*Hufschmied* vol. 52, 1934, 81—84); FRASER, A. D., *Recent Light on the Roman horseshoe* (*Class. J.* vol. 29, 1933/34, 689—691); MÉAUTIS, G., *Les Romains connaissaient-ils le fer à cheval* (*Rev. étud. anc.* vol. 36, 1934, 88); PANNICKE, W., *Auf welchen Grundsatzen beruhen die verschiedenen National-beschlage* (*Hufschmied* vol. 53, 1935, 157); LEFEBVRE DES NOËTTES, *La question du fer à cheval* (*BSAF* 1936, 76—82); CARNAT, G., *Le fer à cheval à travers l'histoire* (Paris, 1951); CARNAT, G., *Das Hufeisen in seiner Bedeutung für Kultur und Zivilisation* (Zúrich, 1953); ARMAND-CALLIAT, L., *Les origines de la ferrure à clous* (*Rev. archéol. de l'Est* III, 1952, 32—36); KOSSACK, G., *Pferdegeschirr aus Gräbern der ältern Hallstattzeit Bayerns* (*Jarhrb. Z. Museum* Mainz I, 1954, 111—178); LEBEL, L. P., *La ferrure à clous des chevaux* (*Rev. archéol. de l'Est* III, 1952, 178—181)

143. JACOBI, H., *Hatten die Römer Steigbügel* (*Germania* vol. 6, 1922, 88—93); RIETH, A., *Halssporen am Pferdegeschirr des Neuen Reiches* (*Mitt. Orientf.* V, 1957, 148—154; SPRAGUE DE CAMP, L., *Before Stirrups* (Isis, vol. 51, 1960, no. 164, 159—160); STEUBEL, Fr., *Fahren und Reiten im Uebergang vom Altertum zum Mittelalter* (Hamburg, 1954)
144. WEBER, L., *Der Schöne Brunnen* (*Z. f. Dtsches Altertum* vol. 63, 1926, 129)
145. HASKINS, C. H., *Studies in Medieval Culture* (Cambridge (Mass.), 1927)
146. STÖHR, K., *Das Nachrichtenwesen des weströmischen Kulturkreises von der Völkerwandrung bis zum Tode Karls des Grossen* (Halle, 1933)
147. FEUCHTINGER, M. E., *Das Verkehr im Wandel der Zeiten seit dem Jahre* 1000 (Berlin, 1935); STEPHAN, H., *Das Verkehrsleben in Mittelalter* (*Raumers hist. Taschenb.* vol. 9, 1869); JUSSERAND, J. J., *English Wayfaring Life in the Middle Ages* (*XIV. Cy.*) (London, 1931); FORBES, R. J., *Bibliography of Road Building A.D.* 300—1840 (*Roads and Road Construction* 1938, 189—196); QUAST, H., *Denkmäler am Strassenrand* (*Strassenbau* vol. 31, 1940, 119—121; 133—137); HALPHEN, L., *Paris sous les premiers Capétiens* (Paris, 1909); FORBES, R. J., *Land Transport and Roadbuilding* (1000—1900) (Janus XLIV, 1957, 2, 104—140)
148. Patentroll 27 Edw. III in RYMER (ed. 1708), vol. V, 774
149. F. IMBERDIS, *Les routes médiévales* (*Ann. Hist. Soc.* vol. I, 1939, 411—416); ARCO, R. DEL, *Modificaciones de vias en la edad media* (*Archiv. Esp. Arqueol.* XXVII, 1954, 295—300); BUQUET, H., *Quelques précision sur les chaussées anciennes et modernes* (*Rev. Int. Onomast.* V. 1953, 129—130); CRAWFORD, O. G. S., *Useless devices* (*Antiquity* XXIX, 1955, 118—119); OOTEGHEM, J. VAN, *Les routes romaines* (*Les Etud. Class.* 1949, 251—262); HOBLER, CHR., *The Badge of St. James* (*The Scallop, edit. Ian Cox*, London, 1957, 51—60)
150. LAURENT, H., *La draperie des Pays Bas en France* (Paris, 1935, 233)
151. FEUCHTINGER, (*Strassenbau* vol. 27, 1936, 168—173); LUDWIG, F., *Untersuchungen über die Reise- und Marschgeschwindigkeit im XII und XIII Jahrhundert* (Berlin, 1897); BERCHEM, D. VAN, *Du portage au péage, le rôle des cols transalpins dans l'histoire du Valais celtique* (*Museum*

Helveticum XIII, 1956, 199—208); SORRE. M., HAUDRICOURT, A. G. et BRUNHES-DELAMARRE, J., *L'homme et la charrue à travers le monde* (*Ann. Géogr. Fr.* 64, 1955, no. 345, 381)

152. MACKAY, E., *Early Indus civilisations* (London, 1948, 17—20); BRUNN, W. VON, *Eine uralte Wohnungs- und Siedlungs-Kultur in Indien* (*Technikgeschichte* vol. 25, 1936, 144—145); GRAY, H. F., *Sewerage in ancient and medieval times* (*Sewage Works* J. vol. 12, 1940, 939—946)
153. DUTT, B. B., *Townplanning in ancient India* (Calcutta, 1925); MEYER, J. B., *Das Arthacastra des Kautilya* (Leipzig, 1925)
154. MARSHALL, J., *Bhîtā* (*Ann. Rep. Archaeol. Survey India* 1911/12); MARSHALL, J., *Taxila* (*Ann. Rep. Archaeol. Survey* India 1912/13, 1914/15, 1915/16)
155. PHILOSTRATUS, *Vita Apoll.* II. 23
156. *Súkranitisāra* I. 531—532; 585—586
157. REUTHER, O., *Die Innenstadt von Babylon* (Leipzig, 1926); ANDRAE, W., *Das wiedererstandene Assur* (Leipzig, 1938)
158. EGGERS, G., *Wasserversorgungstechnik im Altertum* (*Technikgeschichte* vol. 25, 1936, 1—13); BUFFET, B., et EVRARD, R., *L'eau potable à travers les ages* (Liège, 1951), see also Vol. I, p. 149 of my Studies
159. PENDLEBURY, J., *A Handbook to the Palace of Minos* (London, 1933)
160. ARISTOTLE, *Politics* II. 5
161. WYCHERLEY, R. F., *How the Greeks built Cities* (London, 1949)
162. THUCYDIDES, I.10,2
163. PAUSANIAS, VI.24.2
164. STRABO, XIV. cap. 646
165. ARISTOPHANES, *Acharnians* 616—617
166. ARISTOPHANES, *Wasps* 259 (but see RODNEY S. YOUNG, *An Industrial District of Ancient Athens* (*Hesperia* vol. XX, 1939, 145—167))
167. NIELSEN, H. A., *Die Strassenhygiene im Altertum* (*Archiv f. Hygiene* vol. 43, 1902, 85—115; HUGILL, W. M., *The condition of streets in Ancient Athens and Ancient Rome* (*Class. Weekly* vol. 26, 1932, 162); HARRISON, F. H., *The History of drainage, irrigation, sewage disposal and watersupply* (*Bull. New York Acad. Med.* vol. V, 1929, 887—938)
168. PLATO, *Laws* 763, 779; ARISTOTLE, *Oecon.* II. 5.314
169. XENOPHON, *Ath. Pol.* III. 4
170. VITRUVIUS, I. 6; TACITUS, *Annals* XV. 43
171. *C.I.Gr.* 4240, 3945, 3952
172. PLATO, *Laws* 761; ARISTOTLE, *Pol. Athen.* 189—191
173. *C.I.Gr.* 2644, 2570, 3484, 2782, 5141; OLIVER, J. H., *The date of the Pergamene astynomic law* (*Hesperia* XXIV, 1955, 88—92)
174. DEMAN, E. VON, *The Sullan Forum* (*J. Roman Stud.* 1922. 4)
175. LIVY, X. 27; MOMMSEN, Th., *Zum Römischen Strassenwesen* (*Hermes* vol. 12, 486)
176. LIVY, XXXVIII. 28
177. SUETONIUS, *Claudius* 25; CAPITOLINUS, *Marcus Aurelius* 23
178. *C.I.L.* X. 5807
179. NISSEN, H., *Pompeianische Studien* (Leipzig, 1877, 516); BLAKE, M. E., *The pavements of the Roman buildings of the Republic and early Empire*

(*Mem. Amer. Acad. Rome* vol. 8, 1930, 9—159); NISSEN, H., *Italianische Landeskunde* (Leipzig, 1902, II.1, 49—61)
180. ASHBY, Th., *Practical Engineering in Ancient Rome* (*Nature* 1925, vol. 116, 567—580)
181. PLINY, *Nat. Hist.* 36.24
182. LIEBENAM, W., *Städteverwaltung in römische Kaiserreiche* (Leipzig 1900, 145)
183. *C.I.L.* X.5416; V.8668; V.1008a; X.3851
184. *C.I.L.* X.6954
185. AMMIANUS MARCELLINUS, XIV.1; LIBANIUS, *Antiochia* 266; but see also PROCOPIUS, *Anekd.* 26
186. LOESCHKE, S., *Antike Laternen und Lichthauschen* (*Bonner Jhrb.* vol. 119, 1909, 370—430); ROBINS, F. W., *The story of the lamp and the candle* (London, 1939; NILSSON, M. P., *Lampen und Kerzen im Kult der Antike* (*Opuscula Archaeologica* vol. VI, 96—111); see also my *Studies* vol. VI, page 119 ff.
187. SPANO, G., *La illuminazione delle vie di Pompeii* (*Atti Acad. Napoli* N.S. vol. 7, 1920, 1—128); LAMER, H., *Strassenbeleuchtung in späterem Altertum* (*Ph. W.* vol. 47, 1927, 147)
188. RUINAART, *Acta martyrum* 479; HIERONYMUS, *in Isa.* 672
189. TERTULLIAN, *Apologeticum* 35; *Ad uxorum* II.6; *De idol.* 15
190. *Codex Theodos.* XVI. 10.12
191. HERODOTUS, II.44; AUGUSTINUS, *De civ. dei* XXI.6; LUCIANUS, *De dea Syria* 32
192. PLINY, *Nat. Hist.* 37.5.17
193. HEIL, B., *Die deutschen Städte und Bürger im Mittelalter* (Leipzig, 1921, 104)
194. HOLMES, U. T., *Daily living in the twelfth century* (Madison, 1952); BOUTEVILLE, M. R., *L'éclairage public à Paris* (*Rev. Scient.* vol. 71, 1933, 609—615)
195. SABINE, E. L., *Latrines and cesspools of medieval London* (*Speculum* vol. 9, 1934, 303—320); SABINE, E. L., *City cleaning of medieval London* (*Speculum* vol. 12, 1937, 19—43); HEISCHKEL, E., *Die Strasse in der Geschichte der Medizin* (*Technikgeschichte* vol. 23, 1934, 35); MENGERINGSHAUSEN, M., *Die häusliche Verwendung von Wasser und Abwasserwirtschaft im Mittelalter* (*Technikgeschichte* vol. 25, 1936, 43—56); THORNDIKE, L., *Sanitation, baths and street cleaning in the Middle Ages and the Renaissance* (*Speculum* vol. 3, 1928, 192—203)
196. SALUSBURY, G. T., *Street Life in Medieval England* (Oxford, 1948)
197. BAUDRY DE SAUNIER, L., *Histoire de la locomotion terrestre* (Paris, 1936, vol. II, 70)

Date	Political Events	Roadbuilding
ca 450 B.C.	Begin of colonisation policy	Roads existing (possibly as tracks only): V. Latina, V. Gabina, V. Salaria V. Appia till Capua (Via Norbana) V. Flaminia till Narni V. Cassia and Clodia in Etruria perhaps some stretches paved and some gravel roads.
400 B.C.		Paved streets (limestone slabs) existing in Vetulonia and Artena
390 B.C.		Oldest existing pavement in Rome(?)
ca 350 B.C.	First Samnite War	V. Latina to Mt. Algidus, extended to Liris Valley—Cales—Capua as "via vicina" (gravelroad?)
ca 325 B.C.	Second Samnite War	Lime-mortar introduced through the Hellenistic cities of Magna Graecia
312		Censor Appius Claudius builds the Via Appia to Capua (132 miles)
303—301		V. Valerianus built (Rome—Fucine Lake—Carsioli—Alba)
299		V. Flaminia rebuilt (Rome—Ocriculum Narnia) after colonisation Narnia.
290—289	Third Samnite War	Sidewalks built along Via Appia
296—293		Oldest pavements mentioned in Rome; peperino slabs from temple of Mars to Porta Capena laid by order of aedils. Tufa slab pavement ("silex") from Porta Capena to Bovillae on Via Appia.
280—275	Pyrrhus in Italy	

Date	Political Events	Roadbuilding
270—225		Via Appia extended to Tarentum—Brundisium in several stages
268	Colonisation of Beneventum	
264—240	First Punic War Colonisation of Spoletium	V. Flaminia extended to Spoletium
225		V. Clodia paved by censor Claudius Centho.
225—222	Gallic Wars	
220		V. Flaminia extended to Fanum Fortunae and finished
218—201	Second Punic War	
218	Placentia and Cremona colonised	
217—203	Hannibal in Italy	Period of rest
215—205	First Macedonian War	
210		V. Valeria (Sicily) started
abt. 200		V. Latina rebuilt and paved
200—197	Second Macedonian War War against the Insubres (Gauls)	
190	Placentia and Cremona refounded	Street from Temple of Mars to Campus Martius paved in Rome
189	Colonisation of Bologna	
187		M. Aemilius Lepidus builds V. Aemilia (Ariminium—Bononia—Placentia)

Date	Political Events	Roadbuilding
184		Revision and rebuilding of roads and drains in Rome
181	Aquilea founded	
177	Colonisation of Luna	One of the Aurelii Cottae builds the V. Aurelia (Rome—Luna)
176—175		Branch road V. Aemilia built (Mutina—Aquilea)
174		Pavement and sidewalks in Roman streets becoming common. Paved streets in Pisaurum. Roadbuilding let to contractors by censors
171—168	Third Macedonian War	
171		Old track Rome—Aretium rebuilt as Via Cassia
154—125		Via Cassia extended from Aretium to Florentium and Pisa
149—146	Third Punic War	Via Postumia built (Genua—Placentia—Cremona—Aquilea) V. Aemilia extended (Placentia—Dertona)
146	Destruction of Carthage	
148—132		V. Flaminia extended along Adriatic coast to Aquilea and Brundisium
154—133	Conquest of Spain	
abt 145		V. Egnetia built (Dyrrachium—Apollonia—Thessalonica—Hebrus)
142		First stone bridge across the Tiber

Date	Political Events	Roadbuilding
133—101	Tib. and C. Sempr. Gracchus. Unrest in Italy	
132 B.C.		Old track rebuilt as Via Popillia (Capua—Messina)
abt. 129		M. Acquilius rebuilds roadsystem in Asia Minor
123		*Lex Sempronia Viaria* adopted. New roads, milestones and bridges planned
120		Cn. Dominitius Ahenobarbus builds Via Domitia (Rhône—Pyrenees) V. Gabrina built between Salona and Adretium; Streets in Alatri paved. Road from Pyrenees to Numantia and Gades finished?
117		Consul L. Caec. Metellus builds V. Caecilia (Rome to Hatria) Law of the repairing of roads
113—101	War against Cimbres and Teutones, Consulate of Marius	
109		V. Aemilia Scaurus built (Luna—Genua); Genua—Dertona improved
100		V. Aurelia and Cassia rebuilt as highways; V. Domitia & Egnetia enlarged and rebuilt
88—63	Wars with Mithridates of Pontus	
83—79	Sulla Dictator	

Date	Political Events	Roadbuilding
77		Pompey builds road over Mt. Genèvre (Alpis Cottica) from Arles—Valence—Vienne to Turin
64—63	Pontus and Syria conquered	
60	First triumvirate: Pompey—Caesar—Crassus	
57		Caesar improves road over Great St. Bernard (Martigny—Geneva)
58—51	Conquest of Gaul	
46—44	Caesar dictator perpetuus	V. Augusta begun at Valencià. New law forbidding wheeled traffc in Rome
43	Second triumvirate: Antony, Octavianus, Lepidus	
43—30	Civil wars	
30	Egypt becomes Roman possession	
30 B.C.–14 A.D.	Augustus emperor	
27 B.C.		Augustus repairs part of V. Flaminia; C. Calvisius Sabinus pays for repairs V. Latina
22 B.C.		Foundation of "cura viae" boards
20 B.C.		San Bernardino pass repaired, new road built
15 B.C.		Drusus builds new road over Reschenscheideck between Adige- and Inn valleys
13 B.C.		V. Julia Augusta built (Arles—Genua) over Riviera

Date	Political Events	Roadbuilding
7 B.C.		New road built Hilice—Murcia—Guadix—Cazzlona
8 B.C.		V. Augusta ready (between Tarragona and Sarragossa)
		V. Augusta rebuilt (between Narbonne and Gades)
3 A.D.		V. Sebaste built (Ephesus—Caesarea—Apamea—Iconium)
5 A.D.	Conquest of Raetia and Noricum	M. Vipsanus Agrippa begins new roads in Holland and Lower Rhine districts
9 A.D.	Defeat of Varus	
11—12 A.D.		Road built in Spain between Braga—Tuy and Astorga
14—37	Tiberius emperor	
14	War in Germania	Road in Africa (Gabes—Tebessa)
20		Bribiesca—Pampeluna built in Spain
24		Camino de la Plata built (Merida—Salamanca)
32—33		Road from Braga to Chaves built
37—41	Caligula emperor	
41—54	Claudius emperor	
44		First part of paved Watling Street extended to Verulamium (England). Road Sarragossa—Ilerda—Barcelona—Narbonne built in Spain
abt. 50		English roads extended from Silchester to Bath, Old Sarum and Gloucester, Watling Street extended to Silchester
54—68	Nero emperor	

Date	Political Events	Roadbuilding
68—69	Galba—Otho—Vitellius	
69—79	Vespasian emperor	
75		African road (Bona—Tebessa) built
79—81	Titus emperor	V. Domitiana built (Terracina—Neapolis—Rhegium)
81—96	Domitian emperor	
abt. 85		Sarragossa—Toledo—Merida—Lisbon: built
96—98	Nerva emperor	
98—117	Trajan emperor	
		New road Coptos—Berenice (Egypt)
abt. 100	Greatest extent of Roman Empire	African road from Tebessa—Timgad—Constantine
108—112		Via Trajana built (Beneventum—Canusium—Brundisium)
117—138	Hadrian emperor	
123		New road Tebessa—Carthago built (Africa)
abt. 125		Big repairs of Via Appia

THE REPAIR OF ROADS IN SPAIN DURING THE ROMAN EMPIRE (after VAN SICKLE)

Date	Emperor	Roads						
		Via Augusta Tarragona—Cadix	Via Argenta Merida—Salamanca	Braga—Astorga (via Tuy)	Braga—Chaves	Via Nova Braga—Astorga	Braga—Lissabon	Bribiesca—Bordeaux
32 B.C.—14 A.D.	Augustus	built 8 B.C.		built 11/12 B.C.	built 32/33 B.C.	—	—	—
14 A.D.—37	Tiberius	repair 34	built 24/25	sm. rep. 32/33	—	—	—	built (?A.D.)
37—41	Caligula	sm. repairs	—	—	—	—	—	—
41—54	Claudius	sm. repairs	rep. 45/46 & 50	—	sm. rep. 44	—	—	—
54—68	Nero	,, ,,	sm. rep. 57/59	sm. rep. 55	—	—	—	—
68—69	Galba, Otho, Vitellius	—	—	—	—	—	—	—
69—79	Vespasian	repair 79	—	—	—	built 77/80	—	—
79—81	Titus	—	—	—	—	—	—	—
81—96	Domitian	repair 90	—	—	—	—	—	—
96—98	Nerva	sm. rep. 97	—	—	—	—	—	—
98—117	Trajan	repair 98	repair 101/102	sm. rep. 98	repair 104/105	—	built ca 100	—

Date	Emperor	ROADS						
		Via Augusta Tarragona—Cadix	Via Argenta Merida—Salamanca	Braga—Astorga (via Tuy)	Braga—Chaves	Via Nova Astorga Braga—	Braga—Lissabon	Bribiesca—Bordeaux
117—138	Hadrian	—	sm. rep. 121/122	rebuilt 134	—	rebuilt 134/135	—	—
138—161	Antonius Pius	—	—	—	—	—	—	—
161—180	Marcus Aurelius	—	—	—	—	—	—	—
180—193	Commodus (Pertinax & Julianus)	—	—	—	—	—	—	—
193—211	Severus	—	sm. rep. 200	—	—	—	—	—
211—217	Caracalla	rebuilt 214	rebuilt 214	rebuilt 214	sm. repairs	rebuilt 214	rebuilt 214	—
217—222	Heliogabalus	—	—	—	—	—	—	—
222—235	Alexander Severus	—	—	—	—	—	—	—
235—238	Maximin	sm. rep. 237	sm. rep.	—	rebuilt 238	repairs 238	—	—
238—249	Gallienus, Philippus, Gordan	—	—	rebuilt 249	—	—	—	—

Date	Emperor	Roads						
		Via Augusta Tarragona—Cadix	Via Argenta Merida—Salamanca	Braga—Astorga (via Tuy)	Braga—Chaves	Via Nova Braga—Astorga	Braga—Lissabon	Bribiesca—Bordeaux
249—251	Decius	—	—	—	—	—	—	—
253—260	Valerian	—	—	—	—	—	—	—
260—268	Galienus	—	—	—	—	—	—	—
270—275	Aurelian	sm. rep. 272	—	—	—	—	—	—
275—282	Tacitus	—	—	—	—	—	rep. 276	—
284—305	Diocletian	—	—	sm. rep. 300	sm. rep. 282/283	—	—	—
312—337	Constantin the Great	—	—	—	—	rebuilt 317/323	repair	repair
337—340	Constantin II	—	—	—	sm. repairs	—	,,	,,
360—363	Julian Apostata	—	—	—	—	—	—	—
363—364	Jovian	—	—	—	—	—	—	—
364—375	Valentinian I	rebuilt 364	—	—	—	—	—	—
375—378	Valens	—	—	—	—	—	—	—
378—392	Valentinian II	—	—	—	—	—	—	—
375—383	Gratian	—	—	sm. repairs 380	—	—	—	—

Chapter four

THE COMING OF THE CAMEL

Introduction and natural history

The coming of the camel marks an important stage in the history of land transport, for it enables man for the first time to establish a regular trade and contact across the desert. However, the camel had been the property of the desert dwellers for many a century and the civilisations of the Near East had known him. Here the most widely divergent views have been expressed and this confusion has been heightened because many authors have not distinguished properly between the two-humped Bactrian camel and the one-humped Arabian species which should be called dromedary. Philological arguments used to prove certain views have augmented the confusion for is not the Iranian "ushtra" for camel transferred in Sanskrit to the zebu, and the Greek "elephas" is used in Gothic to denote the camel? It is certain that the Egyptians had no name for the dromedary until very late but that the Accadians knew both species (a).

The modern Semitic terminology of types of dromedaries and camels is very extensive (71). Both species are more or less bound to a desert vegetation of salt-containing plants and herbs. The insects living on these leaves, shrubs and trees have an important influence on the area in which these species can live. Though some older authors have argued that both species were derived from one original animal form, the finds of fossil species of both forms have now definitely proved that each species had its own wild ancestor. Keller (53) was proved wrong when bastards of camels and dromedaries were found to be very strong but barren. Therefore the taming of the camel and the dromedary need not coincide either in space nor in time.

Both animals definitely belong to the pastoral nomads of the desert and the steppes (100) (92). They were domesticated by people whose ancestors had acquired cattle and who were to tame the horse too. Neither camels nor dromedaries are very suitable for warfare and tactical movements. The men who tamed them and who used them as elements of decoration in their art probably soon found out what modern experts tell us. Neither attachement nor habit seems to impress the camel, but he is not wide awake enough to be wild. He remains tamed but by his own stupidity is serviceable to man. Therefore he appears as a most useful animal both with the steppe-dwellers of the northern rim of the ancient oekumene and with the desert-dwellers of Arabia. As long as he is not subjected to large variations of climate he serves men as a beast of burden (mostly the camel) and as a riding animal (mostly dromedaries). Both species were known as such in classical times.

The dromedary, who is at home in Arabia and Northern Africa is most serviceable between the age of five and fifteen, when he can carry loads upto 1000 lbs. With an average load of 500 to 1000 lbs he covers some 25 miles a day (maximum 60 miles) with an average speed of five to seven miles, and three waterless days. He has been known to travel twelve days without water, and does so regularly for periods of five to six days in Somaliland. The Arabs use his milk, his wool, his dung and his hide and call him the "milkgiving palm of the animals". Allah must surely despise the Christians for not having given them the camel and his milk. His haughty look is due to the fact that men know only 99 names of God but he knows the hundreth one.

The two-humped species, the true camel, is at home in the Asiatic deserts from Iran upto the Gobi. Local terminology of the camel is also very extensive and was so in the ancient Iranian language. The camel is somewhat stockier and more hairy. He climbs excellently, whereas the dromedary is of no use in the mountains. Neither animal can not be used on clayey soils in rainy climates. Both types reach maturity at about 17 years but have been known to live upto 40 or 50 years.

The dromedary (Camelus dromedarius) was the southern heat-loving species. Pleistocene fossil remains of this species have been found in southern Europe including Roumania (73) (101) and north Africa (Algiers). Prehistoric remains found at Tell es-Saffije (Palestine) contained bones of dromedaries. Nearly one hundred years ago Horner discussed the remains of fossil dromedaries found at a depth of ten feet near Memphis (Egypt) by HEKEKYAN BEY (62). This prehistoric wild form of the dromedary (Camelus Thomasi Pom.) probably retreated into desert regions before the wetter climate of the Later Palaeolithic. Agatarchides claimed that in his days there still were wild dromedaries in Arabia and they may still be there now. This was the wild form tamed by the desert-dwellers and used as a riding animal long before the introduction of the horse in these regions.

The fossil remains of the wild form of the camel (Camelus bactrianus) have been found in Central Asia, Siberia and even on the banks of the Volga (74). Other remains were excavated on the southern flank of the Himalaya in the Siwalik hills. Wild camels more like the fossil form than like the modern domesticated types were found in the Tarim Basin, near Kukur-nor and Lob-nor by Przewalski and by Sven Hedin. There is therefore no doubt that the camel is a domesticated species, the wild form of which still lives in Central Asia. The use of the camel spread from Central Asia to the Far East and China and also west through Afghanistan and Persia to the Ancient Near East. In fact the classical authors believe it to be a specific Bactrian animal.

The camel must have come early to Bactria and Persia. In Shah Tepe (1) many bones of the Camelus Bactrianus have been found in strata dating back to 3000 B.C. In Anau the knowledge of the camel coincides with the Copper Age. This does not yet imply that the camel was domesticated at this early age, but on the other hand domestication in the steppe zône cannot have been much later than 2000 B.C. Remains of camels in prehistoric settlements in this zône such as a village near Kiev point to this. The camel plays a large part in Iranian religion. The Bundehesch calls the dromedary "an animal that could not climb two mountains", but praises the camel of Bactria, the animal of the blessed. In the Avesta we find many names ending in "-ushtra", that is "possessing camels".

There has been some doubt in the mind of certain authors whether the Avesta refers to camels or dromedaries. Dhalla, in his Zoroastrian Civilisation, stated that the dromedary was first introduced in Persia and that the camel came much later. Though it is true that Mesopotamia and Persia were the border countries between the area of the camel and that of the dromedary, it is almost certain (and supported by archaeological finds) that the camel came to northern Iran first. Not only was this mountainous country more suitable for the use of the camel, but all classical authors agree in calling this region the typical homeland of the camel. The Gaugamela district of Persia is famous for its camels, though "it is wintry and mountainous, and it is on the borders of this portion of Persia that the camel-breeders live" (Strabo, XVI. 737). The camel was probably introduced into Mesopotamia from the region south of the Caspian mountains, where the Aorsi live "who could import on camels the Indian and Babylonian merchandise" (Strabo, XI, 506). The breed owned by these Caspians, was a fine-haired one, valuable for weaving (Ctesias, in *Apollon. Hist. Mirab.* 20).

From the Aorsi the camel passed on west to Asia Minor in the wake of the Persian army. Cyrus used them to frighten the Lydian horses, a story not only told by Herodotus but by many other classical authors (b). The camel was also used by the Goths when they crossed the Danube many centuries later (62). Herodotus also refers to the very swift Bactrian camels (III. 102) which are therefore used to trace the "gold-digging ants". However in Persian times the dromedary had also been introduced as a pack-animal and for army purposes and the camels used by Cyrus in Lydia may well have been dromedaries. Even the Persians themselves confuse these two species and may be the Assyrians did so when they depicted camels as a tribute from Egypt on the Shalmenassar Black Obelisk instead of dromedaries. Dromedaries may therefore well have been used in the Persian desert and trade with India along the southern route (109).

There is no doubt that in Assyrian and Persian times the dromedary was known and used, soon to become a familiar feature of the Near East. The classical authors agree that it is an animal of the desert (Pliny, *Nat. Hist.* VIII, 67.72) and important for carrying loads (Pliny, *Nat. Hist.* XXVIII, 200). Already in Diodorus' time both species were used in Arabia itself. Still the dromedary was the most common form in the Near East and therefore its Semitic name passed into Greek and Latin (c) when it became more popular in Hellenistic times. It was already tamed by the desert-dwellers when the Assyrians kings had to fight Benhaddad and his camel-riders and could demand horses and dromedaries from the Arri (fig. 38) (99). We must now test Philby's statement that the archaeological evidence seems to suggest that the dromedary was a familiar feature of the Middle East including Arabia from early neolithic times and that it is reasonable to assume that its domestication from local forms began in mesolithic times when man was passing from hunting to the pastoral and agricultural stages (86).

The case for Egypt

Apart from the fossil dromedary bones discovered by Hekekyan Bey near Helwan in 1851—1854 we have reports from Dr. Mook on the discovery of "camel-, zebra- and other bones", which in 1878 were just labelled "stone age" and which have never since been subjected to modern criticism (51). We have some definite finds going back to Predynastic times. There is the pottery "camel's head" found by Brunton at Maadi (50) (67), which shows that the camel (or dromedary?)

was then well known and portrayed from life. Nothing similar was found at the contemporary excavation-site of Merimde (50). Secondly there is an ivory tablet bought by SCHIAPARELLI at Luxor from a necropolis north of Gurna, all objects of which are predynastic. SCHWEINFURTH (97) already acknowledged the value of this find that depicts a camel with a man astride, while another man leads the animal with a rope.

There are several finds of early dynastic date. First there is the small vessel of yellow limestone in the form of a recumbent pack camel discovered by MÖLLER at Abusir-el-Meleq and now in the Berlin Museum (68) (88) (95). It dates back to the First Dynasty and this led HALL (37) to claim that the dromedary must have been known already in the Predynastic period. Then there is the pottery camel's head found at Hierakonpolis (10) (82) (83) (87) and two further ones from Abydos, all of First Dynasty date (97). One of the latter is now shown at the Oriental Institute, Chicago (No. 7972). The fayence vessel in the form of a camel mounted by a long-haired Asiatic found by MARIETTE at Abydos seems of New Kingdom date (5), it is now at the Berlin Museum (Gise No. 3830), formerly it was thought to be more ancient (23). Another figurine of a camel (*Fond. Piot* XXV, 247, fig. 10) might also be of First Dynasty date as it was found together with a baboon statuette of the Abydos type.

A three feet long two-strand twist of hair-cord was found in the Fayyum by Miss CATON-THOMPSON (12). On analysis it proved to have been made of camel-hair, thus proving the existence of the camel in Egypt (or Egyptian trade) during the Third Dynasty as there is no doubt about its archaeological context.

SCHWEINFURTH attributes a rock drawing found at Aswan showing a camel, a man and hieratic script, to the Sixth Dynasty, but some doubt this early date (97). Dating rock drawings is always a very difficult task and no certain evidence can be obtained from them regarding our subject. Drawings of camels together with long-horned cattle were reported from rocks east of Etbai (64) and DUNBAR reports on other drawings from the region between Wadi Halfa and Aswan (20). He ascribes the strongly schematised figures of camels to the New Kingdom period.

Figurines of camels at the Leyden Museum may date from the Old Kingdom (68) (69). Though therefore the camel was certainly known in Predynastic and Old Kingdom times we should not yet draw the far-reaching conclusion that the dromedary had been introduced in

Egypt or even in North Africa. We have ample evidence that this was not the case before the Roman Empire as the following data will show.

A Middle Kingdom date can be given to the figurine of a camel in its typical recumbent position as found in Byblos with other Egyptian objects. Then there is the skull of a camel found in the Fayyum by Little which is of Middle Kingdom date, and certainly older than 1400 B.C. Rockdrawings of dromedaries in the Wadi Hammamat (97) (101) also seem to belong to this period.

Several proofs of the existence of the dromedary are of New Kingdom date. There are not only the rock carvings of dromedaries at Gebel Silsileh reported by PETRIE, but also the figurine of a camel found at Médamoud by BISSON DE LA ROQUE. Somewhat younger is the glazed figure of a dromedary with painted water jars found at Benha (5) and a similar pottery figurine found at Rifeh and also dating from the thirteenth century B.C. Finally we have an interesting item mentioned by NASH (72). Among the Egyptian antiquities presented to the Harrow School Museum by the late Sir J. WILKINSON is a steatite seal or stamp engraved with figures of two camels (dromedaries?) standing opposite each other and having apparently a sack or net of forage between them. In Sir J. WILKINSON's catalogue now in the Vaughan library, Harrow this seal is described as the only known Egyptian representation of the camel and is said to have come from Taphis (present Teifa) in Nubia where there is a small temple of Roman date. WIEDEMANN mentions figurines of camels of Aethiopien date in the Museum at Boulacq (105).

We now come to the Assyrian and Persian invasions of Egypt. We believe that FREE (28) should not have cited Shalmanassar's Black Obelisk as a proof of the existence of the camel in Egypt, for this monument not only shows the twohumped camel as coming (possibly) from Egypt, but such lists of booty from distant countries are at most secondary evidence if any. LÉFEBURE (61) quotes the records of the Assyrian king Esarhaddon (681—668) for he did use the "camels of all the kings of Arabia" to enter Egypt through the desert during his tenth campaign. The camel was very useful to the invaders of Egypt for did not Cambyses too use them. "He filled a number of camel's skins with water and loading therewith all the live camels he possessed he drove them into the desert and awaited the coming of the army" (HERODOTUS, III.9). The Persians had them in their army train and later Persian armies contained groups of camel riders "not inferior in fleetness to horsemen" though in battle they had to be kept in reserve for "the horse cannot endure the sight of the camel" (HERODOTUS, VII. 86—87).

Though there is therefore no doubt that the ancient Egyptians knew the dromedary as a fleeting "bird of passage" belonging to the sphere of the desert-dwellers, very much disliked by the Egyptians, this animal was introduced as a beast of burden by the Assyrian and Persian conquests (104). The Persians, who also taught its use to Alexander the Great (QUINTIUS CURTIUS, IV.7.12), thus enlisted the help of the Hellenistic conquerors to introduce the camel into Egypt. In fact the camel and the dromedary seem to have been introduced at the same time though the latter was actually used. The camel remained a curiosity shown drawing chariots at the festival given by Ptolemy Philopator according to ATHENAEUS (*Deipnos.* V.5). LUCIAN describes the awe with which the Alexandrian mob gazed at a black camel during the same festival.

The earlier generations of Egyptians therefore did not breed dromedaries or use them for transport purposes. A country which is traversed by a large river and many canals is primarily interested in water-traffic and only for cross-desert travel would the dromedary be eminently suitable. Some have overstressed the fact that the general type of traffic in Egypt left little place for camel transport (37) (5) (106) (24) (61). This may be true for the Nile valley proper, but from the beginning of history we find the Egyptian government sending its expeditions to Sinai or the Eastern desert using only donkeys as beasts of burden. We must conclude that the dromedary was not yet commonly used or bred in the Nubian desert and that its domestication was confined to the desert dwellers of the Arabian desert, with whom the Egyptians were seldom on friendly terms. Also when they adopted the use of the dromedary it was more or less confined to desert transport where conditions were natural to this kind of traffic. We can not easily explain the fact cited by SELIGMANN (*JEA* III, 127) that the sticks of the camel drivers between Suez, Kene and Koseir still completely show the form of the sceptre of the ancient Egyptian kings except for the forking at the bottom. However it is certain that in Hellenistic times the dromedary was introduced not only for the transport of goods (see the Zenon papyri) but also to carry the less important mail of the cursus publicus (47) and to travel protected by "ephodoi". Therefore its merits were publicly recognized well before Roman times to which certain authors would like to confine its use (91). We know many of the "roads" in the Eastern Desert used by the "ala dromedaria", the camelpatrols, such as the track leading from Kainopolis to Albus Portus by the Mons Claudianos (70). The tracks used by trade can be classed by observing the number of parallel camel tracks which form the "road". VEGETIUS long ago remarked the almost uncanny instinct of the camels for treading in the long obliterated footsteps of their predecessors (46). For these tracks were not paved roads and whereas "in earlier times the cameldrivers used to travel (from Coptos to Myos Hormos) only by night, looking for the stars for guidance, now watering places and cisterns guide them" (STRABO, XVII.315).

According to ERMAN cross-desert trading by camel to Syria did not start until the fourth century B.C. and we have no further data to judge its development. On the other hand we notice that camel-transport spread to the western desert too. Not only were tourists transported to "the pyramids, mounted on camels and keeping the Nile on their right hand" (PHILOSTRATUS, *Vita Apoll.* V. 43), but traffic to the Fayyum was maintained with donkeys and camels. The custom-registers of the Fayyum, which can be traced back to the third century B.C. (93), show that considerable trade passed through the various stations, most of it carried by private owners of donkeys and camels (P. Oslo 53). It may be that the dromedary was introduced by actual migration of Arabian camel-drivers into the Nubian desert, for some papyri speak of the "Sarakènoi" whose home in Arabia Felix proper and who excell in the breeding of camels (DIODOR, III.45).

In Roman Egypt the public camels and donkeys as well as these privately owned were mobilized by the state for the transportation of the annona, which could not be handled by the corporation of private owners alone. We know little about these public camels and the guild of their cameldrivers, but we have data showing that sometimes animals from other nomes were requisitioned. Requisitions of camels were sometimes issued for army needs and for transporting stone from government quaries. The emperor Germanicus had to issue a law forbidding exactions. Declarations of ownership and private sale appear frequently in the records of the second century A.D. and it would appear that the government preferred ownership to rest in private hands.

The private owners of camels and donkeys paid a tax or licence fee called "telos kamelon" of some 10 dr. a head. The owner was held to give the number of his camels and the records of the year and these were carefully checked. Owning camels was not unprofitable for the daily earnings amounted to 1—4 dr. a day. The price of camels varied from 400—800 dr., females being cheaper. It seems that the caravan routes were controlled by private companies, except the important route from Coptos to Myos Hormos, which was run by the state and passes were therefore required from travellers.

Round the beginning of this century a controversy raged when some tried to explain the absence of the camel on ancient Egyptian monuments. Many believed that this was due to the ugliness of the camel (19) or to a religious taboo (61) (13) (41) because the camel was the animal of Seth. The originator of the story of the "unclean camel" was VON KREMER (56). It was soon realised that these speculations were false as new documents and figurines of camels cropped up. May be the Egyptian reliefs never show the Arabs riding and fighting on dromedaries, but the Assyrian monuments do. But then we do not know much of Arabian tactics in those days and may be they dismounted in earlier days before they adopted the harrassing tactics with which they sometimes worried the Assyrians. The controversy was mainly caused by the fact that the Old Testament mentioned camels being given to Abraham (Gen. XII.10) and camels being stricken with other animals by the plagues in Egypt (Exod. IX.13). Possibly these texts represent very much the nomad's view of Egyptian urban life.

A second source of confusion was the fact that CHABAS thought he found traces of the camel in several Egyptian hieroglyphic papyri of the Anastase collection (13) in which he read that the camel was trained to dance. It is now recognised that this (d) refers to a type of monkey and not to the camel, to whom only a late hieroglyphic text refers (a). The strange dance-like movements of the camel were of course observed in Antiquity and even described by HORAPOLLON (II.100): "If they (the Egyptians) want to signify a man whose feet are slow to move they design a camel; for this alone amongst the beasts and animals bends its thigh and that is why it is called kamelos." He seems to believe in the false anatomy given by HERODOTUS, III.103 and AELIAN, X.3, which is, however refuted by ARISTOTLE (*de Anim. Hist. II*.1). His etymology had already been given by ARTEMIDORUS I.4 and it seems to depend on the false analogy of kamelos and kammeros (from kampto, to bend and meros, the thigh).

We can therefore not agree with KELLER that the dromedary was introduced into Egypt by the Hyksos (52). The dromedary was the third beast of burden, following the donkey and the horse (104) in the eastern Mediterranean and penetrating Egypt with the Assyrians to establish itself as a motive factor in Hellenistic times. This at the same time sets certain limits to its introduction in Northern Africa.

The case for Northern Africa

Obermaier and others have rightly maintained that the camel existed in neolithic times in these regions. Some have held that domesticated dromedaries were used by the Libyans in their raids on Egypt but none of the records which go back to about 2000 B.C. proved that these desert-dwellers knew the dromedary (101). It has been rightly claimed that the dromedary was unknown until Roman times in Northern Africa (7) (33). Rock drawings such as those of Abu Sofian (75) in the western desert can hardly be properly dated and may not go back more than 1500 years at most. Neither the Libyans nor the Numidians seem to have possessed camels, they do not play any part in the Jugurthine war. They do not occur in Carthaginian documents and seem to have crept slowly along the coast of northern Africa with the trade caravans (15). We first hear of an appreciable amount of camels when Caesar (*De Bello Africano* 68.4) takes 24 camels at Thapsus, which had only recently been imported by Juba. Synesios (*Epistolae* No. 130) speaks of the use of the dromedary in Cyrene during the reign of Arcadius.

We know that dromedaries played an important part in the wars of the Vandals and later of the Arabs. They were surely in common use long before the Arabs, for the Berbers have their own terminology which is only slowly partly displaced by the Arabian one (62). We have, however no right to conclude from these data that the dromedary was already familiar to the Libyans about 500 B.C. (25). As it came to the Egyptians only some two centuries later and such trips with dromedaries to the Libyan oases as those of Alexander the Great are still recorded as curiosities this can not be correct (4). In fact Libyan camel-riders appear in desert warfare only in the second and third century A.D. (98) when the camel had been definitely introduced there. Its wanderings to the western part of North Africa certainly took another century or two, for it is decidedly incorrect to ascribe its introduction to Septimius Severus, who is said to have wanted to stimulate trans-Saharan trade.

The case for Palestine and Syria

In Syria archaeological finds we prove an early knowledge of the domesticated dromedary (15). Oppenheim found an orthostat (102 L) at Tell Halaf which depicts an armed camelrider. Its date, which Oppenheim put at 3000 B.C. is certainly too high, but it may go well back into the third millennium B.C. and thus proves the high antiquity of the domestication of the dromedary by the desert-dwellers. From them the patriarchs no doubt derived their knowledge of the dromedary (Gen. 12.16; 24,30) but we have few early finds from Palestine too. There is the little figurine from Megiddo, which closely ressembles the First Dynasty Egyptian types. Then camel bones of Middle Kingdom date were found in the Second Semitic Stratum at Gezer. Pre-Canaanite Israel seems to have known the dromedary but like Egypt not to have bred or possessed it. Small cattle (sheep and goats) composed their herds like those of the poor tribes of Arabia until the present day (81).

The Israelites were no true desert-people who could not possibly have upheld the law calling the camel unclean and forbidding to eat its meat (Lev. 11,4; Deut. 14.7). It never became so well established in the early Israelite world as to give rise to any problem. Still the patriarchs are said to have owned them, Abraham (Gen. 24,35; 24,64) and Jacob (Gen. 30,43; 31,17; 32,16). This seems to many a later interpolation for they are completely missing in the story of Joseph, nor do they play any part in the exodus from Egypt. Still they must have been a familiar feature in Palestine as the desert-dwellers come to the borders of Palestine with their herds of camels as part of their seasonal migrations every year.

David had a herd of dromedaries as pack-animals led by an Ishmaelite (1 Chron. 12.47; 27,30) and they are mentioned after the return of Israel from exile in Babylonia (Ezra 2.66). They are constantly mentioned in connection with such desert-dwellers as the Bedouins (Gen. 37.25, 28), the Midianites (Judges 6,5; 7,12; 8,21,26), the Amelakites (I. Sam. 15,3,27,9; 30,17), the king of Saba (I. Ki. 10,2; Chron. 9,1), and those of Elam and Media (Isa. 21,7,9). But then Job possesses them too (Job 1,3; 42,12). Their terminology is taken from the Arabs (e). In the New Testament they are accepted as the largest domesticated animal (Matthew 19,24; Mark. 10,25; Luc. 18,25), its hide and hair are not (or no longer?) considered unclean (Matthew 3,4. Mark. 1,6). Still its grazing ground remained the desert as before (Ezek. 25,5) and the profession of camel-driver long remained synonymous with robber. The steppes of Transjordan and the Negeb were its breeding grounds (3) but its role in warfare becomes increasingly important from our era onwards (21).

THE CASE FOR MESOPOTAMIA

Though the earliest statuettes of dromedaries in this region were found at Warka and date from the Cassite period (thirteenth century B.C.) and a terra-cotta camel head of the fourteenth century B.C. was found at 'Aqar Qûf (43) the animal itself was certainly known earlier for its Sumerian name "ass of the sea-lands" (f) implies that it was introduced by the nomads of the Syrian desert or at least recognized as their special animal. It is not often mentioned in contracts or letters and the probablity is that the Arabs kept their own carrying trade in the desert as a monopoly, rarely showing their beasts in the towns. Camel-caravans, such as now ply between Baghdad and Mosul, either were not common or were distinct from the ordinary methods of travel used by the Babylonians. That camels were not led into the cities is not unusual, as their drivers still prefer to park them outside. It has also been suggested that at this early date (before 1000 B.C.) they were not yet in bred sufficient numbers for supply to the river-lands. Pack-asses were the regular transport-animals in the early kingdoms of the Near East and remained so for a thousand years after the Cappadocian tablets, where they are often mentioned. The desert-dwellers and their dromedaries did not become part of the trade-system of the Ancient Near East (34) for quite some time, though the knowledge of the dromedary and its terminology was transmitted by them to the Mesopotamian world (f) (17).

It seems that the true camel was earlier in common use in Mesopotamia (more precisely in Assyria) than the dromedary which begins to show up more frequently in the texts only by the end of the eighth century B.C. The camel was, however, introduced by the Assyrian king Tiglath Pileser I (about 1100) to Assur, called "ud-ra-a-te". The two-humped camel was then bred and used by Tukulti-Ninurta II and Samshi Adad V. We have mentioned that camels are shown on the Black Obelisk of Shalmanasar III. It has been convincingly suggested that the "Mušri" (Egypt) here stands for the "Mušri" of the Median hills, whence camels had long been introduced as well as from Parsuas (Lake Urmia). This is the region where the classical authors place the breeding of camels too (AELIAN, *Nat. anim.* 17, 33). In the battle of Karkar (853) Shalmanasar III finds an Arab chief Gindibu with 1000 dromedary-riders amongst the enemy army. The bronze reliefs of the same king show camels amongst the tribute given to him by northern Syria and Gilzan.

Tiglath Pileser III boasts to have taken 30.000 dromedaries from the Arabian tribes of Palestine and Syria (745 B.C.). They form an important part of the baggage-train of Sargon II during his eighth campaign, when the "camels and mules jumped like goats in the mountains." Sennacherib during his first campaign takes 5230 camels and later receives them as a tribute from the city of Hararti, from Egypt, Arabia, the Sabaeans, the land Ellipi and from Hezekiah. Essarhaddon gets them from Hazael. They also occur in many of the annals of Assurbanipal. Tiglath Pileser IV and Assurbanipal took so many camels from the Arabs that their price fell from $1^2/_3$ minae to $^1/_2$ shekel.

Both the camel and the dromedary are now in common use in Mesopotamia. When in distress the rider can always cut open its stomach and drink its contents. This is a widely used medicine (see MEISSNER, *OLZ* XIX, 1916, p. 113). The two species are imported from Media and Arabia in great numbers, as they were well suited for transport in the northern steppes of Mesopotamia. In Babylonia they never became too popular, they are mentioned during the reign of Nabupolassar, but the donkey remained the popular beast of burden in this part of Mesopotamia. The British Museum has many terra-cotta figurines of camels from Babylonian and Classical periods and we have a cylinder-seal (26) of the Persian period showing a king spearing a lion from the back of a camel.

It would therefore seem that Assyria played an important part in introducing the camel from Media (about 1100) and the dromedary from Arabia (700 B.C.) using them in their country which was suitable for camel-caravans. In southern Mesopotamia they were generally used at a much later date.

The case for Arabia

The economy of ancient Arabia hinged on the use of the dromedary as a means of transport. Probably the camel was also introduced as early in Arabia as in Assyria, but as the camel is not fitted by Nature to sustain long spells of heat, all trials to acclimatize it in Arabia have petered out. Even nowadays Arabia consists of a few coastal fringes on its north, east and south flanks, about $^1/_6$ of its surface area, which hold $^5/_6$ of its population sustained by some form of agriculture. By far the larger part of Arabia ($^3/_4$) are high grounds in which the desert-dwellers, a quarter of the population, live.

It is now generally agreed that the basis of the early Minyean (1300

B.C.) and Sabaean (800 B.C.) kingdoms of Arabia Felix was camel breeding. Thus the spices, incense and other goods produced there or obtained from India were "shipped" by dromedary caravans to the West. Later other important centres of desert transport such like Petra and Palmyra come to play their part. The classical authors were fully aware of the importance of the dromedary for Arabia. STRABO reports from the west coast of Arabia, that there live "nomads, who get their livelyhood from camels, for they carry war from the back of camels, travel them and subsist upon their milk and flesh" (STRABO, XVI, 777). The "Scenitae keep herds of all kinds particularly of camels" (STRABO, XVI, 765). In *Nabataea* STRABO reports even wild camels (XVI. 777). PLINY tells us of the part they play in the transport of frankincense: "A camel would carry about 500 pounds, hence the cost was then doubled by charges exacted along the route, such charges coming to something over ten denarii per day per camel load for sixty-five days (*Nat. Hist.* XII, 64—65)". DIODORUS (II.54) reports on early camel breeding in Yemen.

The rise of camel-transport as a more international form of traffic began in the Syrian desert (35). It could only become an economic possibility if the two horns of the Fertile Crescent were held by one political power or at least by two friendly ones interdependent on each other and willing to take the short-cut through the dangerous desert to enliven trade and cheapen goods. It is also clear that this must have happened after camel-breeding among the Arabs had provided them with sufficient pack- and riding-animals to start trading their specific products for those of the river-plains and the sea-coast, such as salt on which trade Palmyra grew rich. It was important to have a perennial water-supply along the route, no toll-barriers and the semi-chronic threat of nomad and settled Bedouins firm in hand. It should also be remembered that though the Arabs held camels and had inter-tribal wars, they do not seem to have harassed the inhabitants of Syria and Mesopotamia consistently in the earlier centuries of their contacts as a method of living, though they might contribute warriors to their allies in the Fertile Crescent. The earliest note on professional camel-riding robbers of the desert is given by AMMIANUS MARCELLINUS (XIV, 4 (5)).

After the rise of the kingdoms of Southern Arabia had stimulated the breeding of dromedaries and after camels had been introduced into the Assyrian empire the first possibility for the development of trans-desert traffic came with the Persian Empire. Its centralisation tendencies, a clever combination of decentralisation with certain strongly centralising organisations such as a coinage, postal and road-system in order to wield the heterogeneous masses into one empire are well-known. Tadmor appears in the Assyrian inscriptions of Tiglath Pileser I, but the first traceable roads were built in the Iranian period. These were of course not built in the desert zone itself, but the bridges and other facilities at the approaches of the desert route were made. The real keys to desert-traffic, Palmyra and Petra did not flourish during exactly the same period, because the external political causes of their rise were different. Palmyra flourishing in Iranian times suffered a decline when in the Hellenistic period the direct desert route from Syria to southern Mesopotamia was deserted until well in the second century B.C. Then Palmyra grew in importance again, especially when the Romans and Parthians came to a mutually advantageous understanding on the importance of the desert trade on the eve of the first century A.D. The Roman roads followed the natural routes so ably traced by POIDEBARD and the air forces of France and Great Britain. The Romans developed a series of excellent highways throughout Palestine, Syria and Trans-Jordan, well marked with milestones and paved when advantageous to do so. The desert routes, mostly unpaved but marked out by milestones piled at the sides with pebbles and stones, were often supplied with wells or cisterns at a distance of 30 miles (STRABO, XVI, 748). Palmyrene trade was at its maximum when Petra declined and Armenia became a battlefield between Romans and Parthians.

Palmyra's decline came with its fall in 271, but there was a short revival after the reign of Justinian II, until the Byzantine wars in the early Muslim period. The revival of the eighth century brought commercial prosperity for some five centuries, which dropped off sharply after the sack of Baghdad (1258). A short rise in prosperity ended with the fall of Constantinople (1453). Trade with the west stimulated the desert traffic under the Ottoman sultans in the sixteenth century, but internal disorders in the Ottoman Empire meant a decline until the British dromedary post was opened in 1750 (35).

Petra first became important in the Persian period, and its prosperity rose with the rivalry of the Ptolemies and Seleucids in the second century, which tended to divert all Arabian trade through the desert to Damascus and Syria instead of to the coast or to Egypt where Ptolemy II had tried to draw it. With the conquest of Syria by Pompey Petra's interest in the Palmyra road awakened which remained important until the reign of Trajan. Early in the Christian era Bosra began to take over the role of Petra as the "desert port" of Arabia Felix.

This desert traffic was completely dominated by the dromedary for the horse, though bred by the Arabs for their own use, required relatively careful treatment and more food and water than the dromedary. The domination of Arab life by the dromedary can be studied from the works of many authors such as Musil (71). This domination continued until the motor car invaded the desert in 1923.

Conclusions

Our archaeological data lead us to conclude that the camel was domesticated in Central Asia in late neolithic times. At the same time or perhaps earlier the dromedary was tamed and bred in Arabia from a different species of wild camel. From their contacts with the desert-dwellers the ancient Egyptians knew the dromedary from pre-dynastic times onwards but as the dromedary played no part in their economy it was introduced only in Hellenistic times when trade with Arabia and other countries across the desert began become economically important. Then too the dromedary came to the Nubian desert and the Western Oases, spread to Lybia by the second century A.D. and was adopted by the Berbers some two centuries later. The Jews knew the dromedary from the days of the patriarchs but not being a true desert-people never adopted it. Though they knew it throughout their history it remained "unclean" and played no part in their economy until Roman times. Ancient Mesopotamia knew the dromedary at an early period. About 1100 the camel was introduced from Media for mountain traffic by the Assyrians, who adopted the dromedary for desert traffic about 700 B.C. Trans-desert traffic by camel-caravans, a new feature of land transport, came with the Persian Empire. It was only possible when the whole of the Fertile Crescent was dominated by one or more powers interested in cross-desert traffic as a short-cut.

A point that deserves the attention of future students of this problem is the fact that the dromedary belongs to the Arab-speaking patriarchal groups dominating the desert. The introduction of the dromedary into any other country meant close contact and therefore cultural interchange between these desert-people and the ancient civilisations, a point that has not yet been followed up in more detail (27). Nor do we know much about the extent of the earlier desert trade preceeding the Persian period, which may, however, prove an important link in the interrelationship of the ancient Near Eastern empires, or even of prehistoric groupings.

Notes

(a) Accadian for male camel = ibilu, udra, uduru; for female camel = udrate (see also CAD V, 35ff).
The male dromedary is called gammalu, the female anâkater, both names probably taken from the Semitic terms djamal and nâka.
The Coptic for camel, ϫⲁⲙⲟⲩⲗ, is obviously taken from the Semitic and there is a late hieroglyphic "gml" (P. Berlin 8278, 2/10).

(b) Refer HERODOTUS (I.80); XENOPHON (*Cyrop.* VII.1.48); POLYAENUS (VII.6); AELIAN (*N.H.* III.7).

(c) Dromedary = Accad gammalu, Hebr. gāmāl = Greek kamelos = Latin camelus
The true camel is called (camelus) dromadis by the ancients (from the Greek dromas = fleet-footed).

(d) ERMAN-GRAPOW, *H.W.B.* V.116.10 "kirj", see also V.116.12 "kjkj".

(e) The camel is called gamal, female camels ("bêker") being mentioned only once (Jes. 60.6). Young packcamels are called "bekarim" (Jes. 60.6), riding camels being "kirkârôt" (Jes. 66.20). Caravans of camels are mentioned as "ôrehot" and "halîkôt" (Jes. 21.13; Job 6.19) or "ôrehâ" (Gen. 37.25), much later "šejara" (Ez. 27.25).

(f) The dromedary is called ANŠE (-A)-AB-BA (ibilu) = "ass of the sea-lands", and hence the camel is later called ANŠE-GAM-MAL, gammalu. A queer interchange must have taken place in the early Accadian period, for the earliest animal to be known in Mesopotamia was the dromedary, the Semitic term of which is of course gammalu. The Accadian ibilu is also taken from the Arabic 'ibil.
On the 14th tablet of the ḪAR-RA ḫubullu series the camel is listed as AM-SI-KUR-RA "horned wild ox (elephant) of the mountains". Here the camel and the elephant are ranked as cousins as with the German tribes (ZIMMERN, *Z. f. Assyr.* V, 387). There is a special thistle called AM-SI-ḪAR-RA-AN (Deimel, ŠL No. 70.21) on which camels like to feed.

Bibliography

1. AMSCHLER, J. W., *Die Knochenfunde aus dem Königshügel Shah Tepe, Nordiran* (*FuF* XV. 1939, 115—116); MAXWELL-HYSLOP, K. R., and ZEUNER, F. E., *in Iraq* XVII, 1955, 161—163
2. ANTONIUS, O., *Grundzüge zur Stammesgeschichte der Haustiere* (Jena, 1922)
3. BARROIS, A. G., *Manuel d'archéologie biblique* (Vol. I, Paris, 1939)
4. BASSET, R., *Le nom du chameau chez les berbères* (*Actes Congrès Orientalistes XIV*, Alger, 1905, 69—80)
5. BISSING, F. W., *Die Geschichte des Kamels* (*Ae.Z.* 38, 1900, 68—69)

6. Borchardt, L., *Statuen und Statuetten*, etc. 4 (Berlin, 1934, Blatt 168)
7. Borchardt, P., *Neue Beiträge zur alten Geographie Nord-Afrikas* (*Z.Ges. Erdkunde*, Berlin, 1927, 198)
8. Breasted, J. H., *Geschichte Aegyptens* (Phaidon Verlag, Zürich, 1936, Abb. 38)
9. Wallis Budge, E. A., *Guide to the Babylonian and Assyrian Antiquities* (British Museum, 1922)
10. Capart, J., *Les débuts de l'art en Egypte* (Paris, 1904, 183 & fig. 135)
11. Cary, M., *The Greek World from* 323 *to* 146 *B.C.* (London, 1932, 296)
12. Caton Thompson, G., *The camel in Dynastic Egypt* (*Man* XXXIV, 1934, 24)
13. Chabas, F., *Le chameau chez les Egyptiens* (*Etudes sur l'antiquité historique d'apres les sources égyptiennes et les monuments réputés préhistoriques*, Paris, 1873, chap. VI, 398—420)
14. Charlesworth, M. P., *Traderoutes and commerce of the Roman Empire* (Cambr. Univ. Press, 1924)
15. Cumont, Fr., *Un dieu syrien à dos de chameau* (*Syria* X, 1929, 30—135)
16. Dalman, G., *Arbeit und Sitte in Palästina* (Gütersloh, 1938, Band IV)
17. Delitzsch, Fr., *Assyrische Thiernamen* (*Ass. Studiśn* I, Leipzig, 1874); *The Assyrian Dictionary* vol. V, p. 35ff (Chicago, 1956)
18. Desmolins, E., *Comment la route crée le type social* (Paris, 1927, vol. I)
19. Dornstetter, T., *Abraham. Die Beziehungen zu Aegypten.* III: *Die Geschenke des Pharao an Abraham.* (*Biblische Studien* VII, 1902, 108—123)
20. Dunbar, J. H., *Some Nubian Rock Pictures* (*Sudan Notes and Records* XVII, 1934, 139—167)
21. Dussaud, R., *Les relevées du capitaine Rees dans le désert de Syrie* (*Syria* X, 1929, 144—163)
22. Ebert, K., *Reallexikon der Vorgeschichte* XIV, 109, 213
23. Egger, E., *A quelle époque le chameau a-t-il été introduit en Egypte comme bête de somme?* (*C.R. Acad. Inscr. Belles Lettres* VIII, 1864, 329—330) Brentjes, B., *Das Kamel im alten Orient* (*Klio* 38, 1960, 23—52); Epstein, H., *Le dromadaire dans l'ancien Orient* (*Revue Hist. Sci.* vol. VII, 1954, pags. 247—268); Schauenburg, K., *Die Cameliden im Altertum* (Bonn. Jahrb. CVL—CLVI, 1955—1956, 59—94); Mikesell, Marvin K., *Notes on the dispersal of the dromedary* (*Southwestern J. Anthropology* vol. 11, 1955, 231—245)
24. Ranke, H. & Erman, H., *Aegypten und Aegyptischen Leben im Altertum* (Leipzig, 1923, 586)
25. Flamand, G. B. M., *De l'introduction du chameau dans l'Afrique du Nord* (*Actes XIV Congrès Int. Orientalistes* Alger, 1905, 63—68)
26. Frankfort, H., *Cylinder Seals* (London, 1939, plate XXXVII, m)
27. Frankfort, H., *De archaeologie en de geschiedenis van het Nabije Oosten* (Amsterdam, 1933, 13—14)
28. Free, J. P., *Abraham's camels* (*J. Near East. Studies* III, 1944, 187—193)
29. Froehner, R., *Das Kamel* (*Veter. Histor. Mitt.* 15, 1935, 39)
30. Gadd, C. J., *An Assyrian camp-scene* (*Antiquity* IX, 1935, 209—210)

31. MacDonald, Geo, *Rome in the Middle East* (*Antiquity* VIII, 1934, 373—380)
32. Gandert, O. F., *Haustierfragen* (*Mannus* XXIV, 1932, 372—383)
33. Gautier, A., *Le Sahara* (Paris, 1923, 98); Bovill, E. W., *The Camel and the Garamantes* (*Antiquity* vol. 30, 1956, 19—21); Brogan, O., *The camel in Roman Tripolitania* (P.B.S. Rome XXII, 1954, 126—131)
34. Glanville, S. R. K., *Egyptian theriomorphic vessels in the British Museum* (*JEA* XII, 1926, 58)
35. Phelps Grant, Chr., *The Syrian Desert* (London, 1937)
36. Hahn, Ed., *Die Haustiere und ihre Beziehungen zur Wirtschaft der Menschen* (Leipzig, 1896)
37. Hall, H. R., *Introductory guide to the Egyptian Collections of the British Museum* (London, 1930, 129)
38. Hall, H. R., *General Introductory guide to the Egyptian Collection* (London, 1930)
39. Hartmann, F., *L'agriculture dans l'ancienne Egypte* (Paris, 1923, 212)
40. Hehn, V., *Kulturpflanzen,...* (Leipzig, 1911)
41. Heyer, Herm. Jos., *Bibel und Aegypten* (Münster, 1904, 26—30)
42. Hilzheimer, M., Review of Van Buren, *Ancient Fauna* (*Antiquity* XIV, 1940, 214—215)
43. Hilzheimer, M., *Einige Rätsel aus der Geschichte der Haustiere* (*Z.f. Ethn.* 64, 1932, 139—140)
Albricht, W. F., *Zur Zähmung des Kamels* (*Z.f. Altt. Wiss.* 1950, 315)
44. Hilzheimer, M., *Kamelfragen* (*Naturw. Wochenschr.* N.F. XV, 1916, 460—462)
45. Hilzheimer, M., *Überblick über die Geschichte der Haustierforschung insbesondere der letzten 30 Jahre* (*Zool. Ann.* V, 1913)
46. Hohlwein, M., *Déplacements et tourisme dans l'Egypte romaine* (*CdE* 40, 0940, 253—278)
47. Holmberg, E. J., *Zur Geschichte des cursus publicus* (Uppsala, 1933)
48. Houghton, W., *Was the camel known to the early Egyptians?* (*PSBA* XII, 1889/1890, 81—84; XIII, 1890/1891, 32—33)
49. Huart, Cl. & Delaporte, L., *L'Iran antique* (Paris, 1943)
50. Junker, H., *Vorläufiger Bericht über die Grabungen auf der neolitischen Siedlung von Merimde-Benisalame* (*Anz. Akad. Wiss. Wien*, phil. hist. Kl. vol. 70, 1933, 54—97); Saad, Z. J., *Royal Excavations at Helwan* 1945/1947 (CASAE 14, Cairo, 1951, 38)
51. Keimer, L., *Bemerkungen und Lesefrüchte zur altägyptischen Naturgeschichte* (*Kemi*, II, 1929, 85—90)
52. Keller, C., *Die Abstammung der Haustiere* (Zürich, 1902, 210)
53. Keller, C., *Naturgeschichte der Haustiere* (Berlin, 1905, 190—198)
54. Kennett, R. H., *Ancient Hebrew Social Life and Custom* (London, 1933)
55. Korten, H., *Das Syrische Kamel* (*Das Heilige Land* XXVIII, 1884, 61—64)
56. Kremer, A. von, *Aegypten* (Leipzig, 1863, Band I, 223)
57. Landsberger, B., *Die Fauna des alten Mesopotamiens nach der* 14. *Tafel der Serie ḪARRA*: *ḫubullu* (Leipzig, 1934, 71, 141)
58. Leemans, C., *Monuments égyptiens* (Leyden, 1842, p. 19, Pl. XXIII)

59. Leemans, C., *Aegyptische monumenten van het Nederlands Museum van Oudheden te Leiden* (Leyden, 1867, Deel I, P. XXVII, No. 1019)
60. Lefebre des Noettes, G., *l'Attelage, le cheval de selle à travers les ages* (Paris, 1931)
61. Lefebre, E., *Le chameau en Egypte (Actes XIV Congrès Int. des Orientalistes*, Alger, vol. II, 1907, 24—62); Préaux, Cl., *Vente de deux chamelles (P. Brooklyn gr.* 3) (*CdE* XXXVII, 1962, 37, 155—162)
62. Lehmann, O., *Das Kamel, seine geographische Verbreitung und die Bedingungen seines Vorkommens (Z.f. wiss. Geographie* VIII, 93)
63. Lepsius, R., *L'Etbaye aris* (Berlin, 1854, 101, 161)
64. Linant de Bellefonds, F., *Briefe aus Egypten* (Berlin, 1852, 133)
65. Meissner, Br., *Babylonien und Assyrien* (2 vols. Heidelberg, 1920/25)
66. Meissner, Br., *Akklimatisationsversuche mesopotamischer Fürsten (Assyr. Studien* V, *MVAG* XV, 1910, 5, 3—28)
67. Menghin, O. & Amer, M., *Preliminary report (First season) on Maadi* (Cairo, 1932)
68. Möller, G., *Ausgrabung der Deutschen Orient Ges. auf dem vorgeschichtlichen Friedhof bei Abusir el Meleq (MDOG*, 1906)
69. Mook, H., *Aegyptens vormetallische Zeit* (Würzburg, 1880, 13—14)
70. Murray, G. W., *The Roman roads and stations in the eastern desert of Egypt (JEA* XI, 1925, 138—150); id., *Early camels in Egypt (Bull. Inst. Fouad I du Désert*, Heliopolis, vol. II, 1952, 105—106)
71. Musil, A., *The Rwala Bedouins* (New York, 1928, chapter XI, 338, 348, 357)
71a. Naquib, G., *The Camel in Ancient Egypt (Brit. Veter. J.* London, vol. 106, 1950, 2, pags. 76—81)
72. Nash, W. L., *An Egyptian representation of the camel (PSBA* 24, 1902, 309)
73. Nehring, A., *Fossile Kamele aus Rumänien und die pleistozäne Steppenzeit Mitteleuropas* (Globus 79, 1901, 264—267)
74. Nehring, A., *Ein fossiler Kamelschädel von Sarepta an der Wolga (C. Knoblochi) (Sitzber. Ges. Naturf. Freunde Berlin*, 1901, 137—144)
75. Newbold, D., *Rockpictures and archaeology in the Lybian Desert (Antiquity* II, 1928, 261—292)
76. Noth, M., *Die Welt des Alten Testaments* (Berlin, 1940)
77. Olmstead, A. T., *History of Palestine and Syria* (London, 1931)
78. Pauly-Wissowa, *Realencyclopaedie* XX, 1824—1829 (Stuttgart, 1919)
80. Peake, H., *Early steps in human culture* (London, 1935, 99)
81. Pedersen, Joh., *Israel, its life and culture* (London I/II, 1926; III/IV, 1940)
82. Flinders Petrie, W. M., *Abydos* (London, 1903, vol. II, 27,49 & Pl. X, 224)
83. Flinders Petrie, W. M., *Ranas Ehmasya (Herakleopolis Magnum*, 1904) (London, 1905, 153—154, Pl. LII A)
84. Flinders Petrie, W. M., *Gizeh and Rifeh* (London, 1907, pl. XXVII)
85. Flinders Petrie, W. M., *Wisdom of the Egyptians* (London, 1940)
86. Philby, H. St. J., *Note on Kilwa, Transjordania (Geogr. J.* 102, 1943, 274—277)

87. Quibbell, J. E., *Hierakonpolis* (London, 1902, vol. II, 62,2 & 49)
88. Ranke, H., *Altägyptische Tierbilder* (München, 1925, Abb. 2)
89. Reinhardt, L., *Kulturgeschichte der Nutztiere* (München, 1912, 212—221,
90. Robinson, A. E., *The camel in antiquity* (*Sudan Notes and Records* XIX, 1936, 47—69)
91. Rodd, F. R., *People of the Veil* (London, 1926, 206)
92. Rostovtzeff, M. I., *The animal style in South Russia and China* (*Princeton Monographs* XIV, 1929)
93. Schaal, H., *Vom Tauschhandel zum Welthandel* (Berlin, 1931)
94. Scharff, A., *Grundzüge der aegyptischen Vorgeschichte* (Leipzig, 1927, 43, note 6)
95. Scharff, A., *Altertümer der ägyptischen Vor- und Frühzeit* Band I; *Werkzeuge und Waffen* (Berlin, 1931)
96. Schubart, W., *Aegypten von Alexander den Grossen bis auf Mohammed* (Berlin, 1922, 219—223)
97. Schweinfurth, G., *Über alte Tierbilder und Felsinschriften bei Assuan* (*Z.f. Ethn.* 44, 1912, 627—658)
98. Staffe, A., *Die Haustiere auf den nordafrikanischen Felsbildern* (*FuF* XV, 1939, 344—345)
99. Thompson, R. C., *The Prisms of Essarhaddon and Assurbanipal III* (London, 1910, 46—52)
100. Thurnwaldt, R., *Die menschliche Gesellschaft* (Berlin, 1931, vol. I, 203)
101. Uhden, R., *Zur Geschichte des Kamels in Nord Afrika* (*Petermanns Mitt.* 75, 1929, 307)
102. Vigoureux, A., *La Bible*, etc... *Dictionaire* (Paris, 1912, 406—408)
102a. Walz, R., *Zum Problem des Zeitpunktes der Domestikation der altweltlichen Kameliden* (*ZDMG* vol. 101, N.F. vol. 26, 1951, pags. 29—51; vol. zg. 1954, pags. 45—87); see also the notes of Pohl, A., on these essays (Orientalia 21, 1952, 373 & 23, 1954, 453)
103. Westermann, W. L., *On inland transportation and communication in Antiquity* (*Class. J.* 24, 1928/29, 483—497)
104. Westermann, W. L., *On inland transportation and communication in Antiquity* (*Polit. Sci. Quart.* 43, 1928, 364—387)
105. Wiedemann, A., *Stela at Freiburg in Baden* (*PSBA* XIII, 1891, 32—33)
106. Wiedemann, A., *Varia* (*Sphinx* XVIII, 1914/15, 174—178)
107. Wiedemann, A., *Das alte Aegypten* (Heidelberg, 1920, 198)
108. Wissmann, H. von, *Arabien, Bauern- und Beduinenland* (*FuF* XVIII, 1942, 219—221)
109. Gode, P. K., *Notes on the history of the camel in India between BC 500 and AD 800* (*Janus* XLVII, 1958, 133—138)

Index